Strange
Encounters

BOOKS BY DAVID BEATY
Call Me Captain
The Complete Sky-traveller
Cone of Silence
Electric Train
Excellency
The Gun Garden
The Heart of the Storm
The Human Factor in Aircraft Accidents
Milk and Honey: Travails with a Donkey
The Proving Flight
The Siren Song
Strange Encounters
Sword of Honour
The Take Off
The Temple Tree
The Water Jump: The Story of Transatlantic Flight
The White Sea-bird
The Wind Off the Sea
Wings of the Morning (with Beatty Beaty)

STRANGE ENCOUNTERS
Mysteries of the Air

David Beaty

Atheneum   New York   1984

For B, as always

Contents

List of Illustrations   8

1   The View Through the Keyhole   11

2   Beyond the Inquiry   20

3   Through the Looking-glass   52

4   Not in Our Stars But in Ourselves   60

5   The Other Side of Luck   79

6   The Arms of Coincidence   95

7   The Sixth Sense   103

8   The Frame of Our Destiny   121

9   The Time Machine   133

10   The Rest of the Room   142

Bibliography   152

Index   154

List of Illustrations

between pages 96–97

1a   The York aircraft *(Popperfoto)*
1b   The R101 at its launching mast *(Illustrated London News)*
2a   Inside the R101's control cabin *(Daily Mail)*
2b   Squadron Leader E.L. Johnson, Sir Sefton Brancker, Lord Thomson and Lt Col. V. Richmond, all of whom died in the R101 *(Daily Mail)*
3a   Flight Lieutenant H.C. Irwin *(BBC Hulton Picture Library)*
3b   Eileen Garrett *(Keystone Press Agency)*
3c   Air Chief Marshal Lord Dowding *(Keystone Press Agency)*
3d   Muriel, Lady Dowding *(Lady Dowding)*
4a   An aerial view of the R101's wreck *(Daily Mail)*
4b   The French poacher Alfred Rabouille giving evidence at the R101 inquiry. The model shows the airship's angle at the time of the crash. *(Daily Mail)*
5a   The *Hindenburg* bursts into flames *(Daily Mail)*
5b   A B314 flying boat of the type in which Churchill flew across the Atlantic *(Pan American)*
6a   Amelia Earhart *(Keystone Press Agency)*
6b   Amy Johnson *(Popperfoto)*
6c   Amelia Earhart, photographed from the navigator's seat, at the controls of the plane in which she and Fred Noonan disappeared *(Daily Mail)*
7a   Amy Johnson and her husband Jim Mollison *(Flight)*
7b   Kingsford Smith (right) and Ulm (left) on arrival at Croydon from Australia *(Flight)*
8a   Charles Kingsford Smith in *Lady Southern Cross* over Oakland, California at the end of his flight across the Pacific *(Daily Mail)*
8b   The last photograph of Kingsford Smith, taken as he left England for Australia *(Daily Mail)*
9a   A Boeing Stratocruiser *(Popperfoto)*
9b   Captain H. Gulbransen *(Captain Gulbransen)*
9c   Dag Hämmarskjold *(Popperfoto)*

10a   General Tshombe *(Popperfoto)*
10b   General Sikorski *(Popperfoto)*
10c   A Liberator *(British Airways)*
11a   A De Havilland Comet I *(Flight)*
11b   Captain Harry Foote *(Mrs Chris Foote)*
12a   Captain Norman Macmillan *(BBC Hulton Picture Library)*
12b   The Fairey Postal Monoplane *(Flight)*
12c   Flight Lieutenant H.H. Jenkins and Squadron Leader A.G. Jones-Williams *(Daily Mail)*

Foreword

As usual with every book, the author owes much to more people than there is space to name – those who have talked to him and those whose books he has read.

But I would especially like to thank Fred Browdie, Eileen Coly, Captain Val Croft, Lady Dowding, Mrs Chris Foote, T.N.C. Garfit, Sir Victor Goddard, Elizabeth Grey, Captain H. Gulbransen, Captain Jim Howard, Archie Jarman, the late Captain Kelly-Rogers and Mrs Kelly-Rogers, Mrs Locke-Wheaton, Mrs Stiles, the late Major Villiers, Bruce Waugh, A. Wesencraft, and many relatives and friends of the crew of the R101 who kindly wrote to me and answered my queries.

I have been particularly fortunate in my editors, Ann Mansbridge and Cathryn Game, and I would also like to thank Bob Dignam of the *Daily Mail* Picture Library, the staff of the Public Records Office, the librarians of the Royal Aeronautical Society, the Royal Air Force and Civil Aviation Authority, the College of Psychic Studies, and the Central Newspaper Library for their unfailing helpfulness.

1

The View
Through the Keyhole

Just before nine o'clock on 1 February 1953 five men and one woman arrived in cars at Stansted airport, thirty miles north of London.

The day was cold and grey, sleeting with rain. The woman disappeared into the Catering Section, three of the men walked over to Skyways Operations, the other two went out to the tarmac and on to a silver-sprayed York where they began routine inspections on radio and engines respectively.

G-AHFA was an ugly aircraft with a high wing and four jutting in-line engines that seemed to squat on two fat main wheels and a tiny tail wheel. The previous day, in the process of being towed tail-first across the runway to calibrate its direction-finding loop, a powerful gust of wind had caught the elevators, depressing them fully with such violence that both control columns struck the blind-flying panels, smashing several instruments. As a result, both blind-flying panels were changed, the instruments tested, and an unlicensed airframe inspector made a thorough visual check of the complete elevator control run and saw that all controls were fully serviceable.

The next day the aircraft received its usual daily inspection and was signed up and refuelled. Captain Nicholls, First Officer Walton and Navigating Officer Chopping completed their flight plan to Jamaica via Lagens in the Azores and Gander in Newfoundland. The zigzag route was necessary because the York had insufficient range to fly the 2000-mile overwater leg from Lagens to Bermuda. Half an hour later buses brought thirty-three service passengers to the terminal for a flight carried out under contract with the Air Ministry. G-AHFA was loaded up. The captain and crew came on board. The engines were started. The chocks were moved away. At 11.06 GMT the York lumbered down the runway and lifted off into the wintry air. The flight to Lagens was uneventful, and Captain Nicholls landed at 19.13.

Then there was a mechanical delay. Number 3 engine (the starboard inner) was vibrating and six sparking plugs had to be changed. How the fuel was put into the fuel tanks is not known, but the engineer officer signed for 2085 gallons of aviation gasoline and probably checked the quantity. He also removed the water tank caps which are situated several feet aft of the wing centre fuel tank and were clearly marked as well as being distinct from the fuel

11

tank caps in size and general appearance. Fifty or sixty litres of water were pumped into the tank, after which the engineer replaced and screwed the tank caps.

At last, after four hours and twelve minutes on the ground, G-AHFA took off at 23.25 down the lighted flarepath and set course for Gander at 8000 feet.

The flight proceeded normally. Position reports were received regularly on the hour, showing the York was bucking a 25-knot headwind which reduced her groundspeed to 160 knots. But the weather was good and G-AHFA was flying above cloud.

No position was sent at 05.00 – it was not unusual for reports to be late – but then suddenly at 05.31 the radio operator at Gander received an XXX (an urgency signal) followed by the position 46° 15′ north, 46° 31′ west. This was immediately followed by 'SOS, SOS, SOS de G-A – '

Then silence. Nothing more was heard.

What had happened?

The subsequent inquiry was to emphasize that the possibilities were 'almost unlimited'. Those that were considered were basically the physical factors of weather, mechanical and human error, though the assessors did just touch on the psychological side. But the limitless possibilities existing in psychology and stretching into parapsychology and beyond were never for a moment considered.

This book is largely concerned with those nebulous reasons which the inquiry did not touch on – made up of ideas, dreams, sudden insights and beliefs, leading through Time and Space. It is along these ghostly stepping stones that we can hope eventually to find the glimmerings of understanding of what life on this planet may mean, and the pressures and influences, still largely unknown, that may affect us.

Let us first look at the possibilities the inquiry *did* consider. The weather as a cause was – quite rightly – dismissed. Though hurricanes erupting from the Caribbean Sea curve across the York's track, this was not the hurricane season. The aircraft was flying in a correctly forecast ridge of high pressure. There was no risk of ice, lightning strike or dangerous turbulence.

Was it mechanical trouble? The aircraft had been properly certificated and serviced. Being made mostly from parts of the Lancaster, Britain's best bomber, it was strong and had not been subject to cabin pressurization. The four Rolls-Royce engines were usually reliable. It is hard to believe that a mechanical emergency could occur so suddenly and catastrophically as to interrupt that SOS. The York could easily maintain height on three engines. Nevertheless a sudden and hitherto not understood technical fault cannot be totally ruled out. Two Tudors, *Star Tiger* and *Star Ariel*, had already disappeared en route to Jamaica, and many technical theories were considered, including a possible malfunction of the autopilot. Flying mysteries can sometimes be later explained by the advance of technology. In February 1958 a BEA Elizabethan crashed taking off from a slushy runway at Munich, killing twenty-three people, including eight Manchester United football players. At the subsequent inquiry the verdict of 'pilot error' was returned against Captain Thain.

The captain's wife was a chemist. Working day and night to clear her husband, she did experiments in her kitchen, and eventually showed that even half an inch of slush on the runway would increase the take-off run of tricycle undercarriage aircraft by forty per cent. Under the conditions at Munich that day, Captain Thain could not have got off – and his name was cleared by the British authorities, though not the German.

But an unknown technical fault developing so suddenly on such a well-tried aircraft as the York is unlikely. What might account for that interrupted SOS was a bomb on board. But this was long before the era of the skyjacker. And in the event of a bomb, the radio officer would have immediately sent SOS, and not started off with the mild emergency prefix XXX.

The captain and crew were experienced and properly certificated. A physiological cause – aviation medicine had long been studied – might have been a factor in that a crew member might have had a heart-attack. But had this been the reason for the interrupted SOS, someone else would have taken over the radio and continued transmitting.

The inquiry, however, noted that the crew had been on duty for over nineteen hours at the time of the distress signal. There were at that time no flight time limitations and the crew might therefore have been slower than usual in reacting to an emergency and the more liable to make a mistake. Confusion between the water and fuel tank caps in the darkness was just possible (such human failure errors were to occur again and again in aircraft accidents over the years ahead) and water might have been put into the fuel tanks under the influence of fatigue.

The court made a recommendation that crew fatigue should receive study 'at an impressive level' – one of the very first times that a psychological factor had been brought into an aircraft accident inquiry. Those were the days when the pilot automatically collected the blame for an accident, once the weather and the aircraft had been exonerated.

After that inquiry, psychological causes gradually began to be studied in relation to aircraft accidents and were occasionally brought into reports. In spite of objections to lack of a scientific basis, recently psychology and the human factor has become respectable, if not completely trusted. Just as flying itself was considered impossible and lunatic – on the turn of the century, *The Times* had offered a prize of ten million pounds to anyone who could fly across London – so psychology had to win an academic chair of respectability before it was considered that it might hold some clues worth exploring further.

In the same connection its cousin, parapsychology, is still shunned in spite of the fact that there are now parapsychology departments in universities. The only accident concerning which parapsychology was ever remotely considered was the 1930 R101 inquiry.

Yet there may be many forces that may have an effect on us and what we do. And it is in the high peaks of sudden death and tragedy that clues on the nature of such forces may best be found. The fact that they cannot be proved and cannot yet be understood does not mean that they should be totally dismissed. Many of such forces have been part of the human inheritance for several thousand years, surviving in beliefs, faiths, superstitions, instincts, habits and archetypes. They are like rocks lodged in human awareness – why?

First, astrology – usually a subject for scorn. Yet it is highly regarded in the east, and few marriages or important occasions are undertaken without a consultation on astronomical events, particularly the position of the sun, moon and planets. In the west, many newspapers and almost all women's magazines carry horoscope predictions. The time and date of birth is considered crucial in forecasting a person's future. The well-known psychologist Eysenck admits that for unknown reasons it seems to work. And in connection with aircraft accidents such as that of the York, it is interesting that the highly respected airline Swissair some years ago instituted an inquiry into the dates and times of birth of their captains. These were compared with those of airline captains who had had accidents, and individual time scales were drawn up to show possible danger periods for Swissair captains, during which they were kept off flying. Psychological experiments have shown that accident-proneness is a phase phenomenon. Some psychologists explain this as a result of bio-rhythms. For most of the time, an accident-prone person is as safe as anyone else. And for hundreds of years the effect of the moon on human and animal behaviour has been recognized almost as certainly as its effects on the tides. Such words as lunatic, such behaviour as moon-madness and baying at the moon indicate its hypnotic effect, a state almost of 'possession' by the moon.

There have been a number of cases of pilots being 'possessed' or hypnotized by some idea – the aircraft's controls are locked, or it is not developing enough speed to get off the ground, there is something the matter with it – and taking action that results in a crash, after which it is discovered that there is nothing wrong with the aircraft at all. Pilots have been hypnotized by revolving lights, by one aspect of an instrument such as the localizer needle of the instrument landing system, ignoring the glide slope and undershooting the runway by as much as twelve miles. Moreover this state of 'possession' can be passed on. The captain of a One-Eleven was 'possessed' of the idea that his aircraft was in a stable stall, which he passed on to his first officer, and together they made a crash-landing, after which it was discovered they had never been in a stall at all.

Everyone nowadays recognizes hypnotism but no one knows how it works. How are the ideas transmitted? Is it by will power? By imagination? By faith? Can people actually move things by the mind? Those believing in psychokinesis believe they do. And a leading neurophysiologist, Dr Grey Walter, attached electrodes to the frontal cortex to transmit electrical impulses through an amplifier to a button which switched on the television. Can a similar influence affect the fall of dice, an ability some people say they possess?

Is there then such a thing as luck? Luck has always been recognized as a quality some people possess. Napoleon chose his generals for their luck. In his book, *Three Dimensioned Darkness*, an experienced British captain called Lincoln Lee says that though he hesitates to say it, in assessing a good pilot, consideration should be taken to his luck. Some pilots I have known have been extraordinarily unlucky for no reason at all – they were all superb pilots. Yet if anything untoward occurred on the fleet – engine trouble, ice, sudden fog – it usually happened to them, while some other pilots well below them in ability sailed through their careers with no unpleasant incidents at all.

But is this luck or preordained? Some scientists tell us that the programme of our lives is stamped on our individual DNA pattern. We take on our future like our skins and little can be changed. Like animals, we inherit instincts for survival and the pursuit of happiness, together with some aspects of knowledge – though less apparently than fishes with their extraordinarily complicated courtship behaviour and birds with their innate ability to fly and to navigate for thousands of miles.

And then there is Fate itself. It appears just as immovable as the programmed pattern produced by an individual's DNA. There do in fact seem to be people like Churchill with an inescapable role to play in our society who seem fated to assume that particular responsibility, often recognizing it before it happens to them. If events are preordained, can they be prophesied? Is there such a gift as clairvoyance? And if so, like hypnotism, how does it work? In classical times great weight and reverence was given to the oracle at Delphi. And though its pronouncements were obscure, even today the atmosphere of the grotto is strangely compulsive. In the sixteenth century Nostradamus produced a book of rhymed prophecies, many of which have been interpreted as having come true. Is this clairvoyance or the power of imagination? For the imagination appears to have the capacity to leap time and space to future events.

There is the mystery of Morgan Robertson's novel *Futility* written in 1898. Its plot concerns a new unsinkable steamship of 75,000 tons called *Titan* that hits an iceberg in the Atlantic at a speed of 24 knots on her maiden voyage during April and sinks with great loss of life. The 66,000 ton *Titanic*, also claimed unsinkable, also on her maiden voyage, hit an iceberg in the Atlantic in 1912 at a speed of 23 knots. *Titan*'s length was 800 feet against *Titanic*'s 882.5 feet. The *Titanic*'s twenty lifeboats (four fewer than *Titan*'s) were quite insufficient for her 2207 souls on board, fifteen hundred of whom were drowned. The assassination of Kennedy, the crash of the R101, earthquakes, volcanic eruptions were seen before they happened, as though disasters by their very magnitude make some kind of impact in Time before they happen.

Then Time itself is as mysterious as Space. Are there different worlds operating at the same time as we are but which we cannot see, as some people believe? Since precognitive glimpses happen so often – like Sir Victor Goddard's brief vision of a future wartime airfield – does this indicate a face of Time other than the one we understand from the clock and the calendar?

Or is all this simply coincidence? But coincidence itself, when examined more closely, does appear to have mysteries of its own. What are called coincidences happen oddly and so frequently that many appear beyond the mathematical laws of chance.

Telepathy is a case in point. Telepathic communication has occurred so often that it is grudgingly admitted to exist, though it still remains a mystery. The expression was coined by the classical scholar Myers, from the Greek for 'feeling at a distance'. Bechterer coined the expression 'biological radio', and certainly thought waves do seem to exist, which presupposes some kind of sending and receiving apparatus.

The simplest example of telepathy is knowing what someone is thinking without them opening their mouths – many husbands and wives have this ability to an uncanny degree. Habit, reading it on the face, or predictable expectation are the usual reasons given to explain this phenomenon, but frequently such communications are sent over considerable distances from one person to another, neither of whom know each other.

In 1932 the psychologist J.B. Rhine began a series of numerous experiments on transmitting and receiving messages. Cards were used with five markings – circle, cross, square, star, wave – and a sender picked them out of a sorted pack and 'transmitted' them to the receiver in another room. Results were such as to rule out chance, and Rhine and his wife founded the first Parapsychology Department at Duke University.

Various experiments in 'transmitting' messages by 'thought waves' have been carried out between the shore and a US nuclear submarine with reported excellent results. Card experiments were carried out with astronauts in Apollo 14, and again the results were reported as being beyond expectations. The Russians have done a series of successful experiments along the same lines between towns hundreds of miles apart. Large programmes investigating the powers of the mind are operating in Leningrad and a number of other cities. The Russians take the research very seriously.

There are of course occasions when the message is wrong. Further investigation has shown that often in such cases, boredom and fatigue sets in. Without motivation, the results descend towards random. Though we may have some telepathic powers, in some areas we appear to have lost our sensitivity. We cannot hear the high notes a dog can, nor pick up a scent over a distance. In certain respects, we see, hear, sense and understand *less* than birds and animals, yet we rank ourselves as the all-knowledgeable creature, believing that what we cannot see, or hear, or sense does not exist.

Yet in fact we communicate minimally, and a case could certainly be made to show that talking has been a retrograde step in understanding. We certainly misunderstand what people tell us. There have been numerous accidents caused by misunderstandings between controllers on the ground and pilots in the air. Since the advent of cockpit voice recorders which tape everything that is said on the flight deck, there have been cases of the captain particularly not getting the message from his first officer, even though we can read in his words what the man was trying to say to him.

In a Convair flying through thick cloud over the Rocky Mountains of the central USA, the first officer had just made an uneasy joke to the captain, 'Well, we must be somewhere in Oklahoma.' The captain had told him to 'descend to two thousand'.

Captain:	We just kinda turned a little bit while you was looking at the map.
First Officer:	Look.
Captain:	First time I've ever made a mistake in my life.
First officer:	I'll be *[1]. Man, I wish I knew where we were so we'd have some idea of the general terrain around this * place.
Captain:	I know what it is.
First Officer:	What?
Captain:	That the highest point out here is about twelve hundred feet.
First officer:	That right?

1 What the typed read-out calls a 'non-pertinent word'.

Whistling starts on the flight deck.

First officer: Two hundred and fifty, we're about to pass over Page VOR[2]
                 . . . you know where that is?
Captain: Yeah.
First officer: All right.
Captain: About a hundred and eighty degrees to Texarkana.
First officer: About a hundred and fifty two . . . minimum on route
                 altitude here is forty-four hundred –

The sound of impact interrupted the tape as the Convair hit a mountain at a height of 2025 feet. All on board were killed. From that short excerpt, we can immediately appreciate that the first officer realized their danger. The mystery is why the captain beside him did not, and what pressures prevented the first officer from simply taking over and climbing. And this is no isolated incident. Since the war, 612 large civil airliners have crashed into mountains, many in not dissimilar circumstances.

We would say that such captains of aircraft 'had no imagination', condemning such lack of forward thinking to a lack of the same quality we use in pejorative sense to brush aside all that we do not immediately understand – telepathy, precognition, unidentified flying objects and other mysterious occurrences in the sky. The first officer on the other hand, 'had imagination'. Certainly in many cases imagination appears to have been involved, but it rarely explains everything, and the central thing it does not explain is what imagination is in the first place.

Do the 'pictures' of our imagination come from inside ourselves inherited from personal memory? Or are they inherited through race memory or from some substance akin to DNA? Or do we pluck them out of what Jung called the Collective Unconscious, a mass of thoughts and feelings of people living and dead? The medium Rosemary Brown writes music which she says is 'received' from great dead musicians. There was a lady in the West Country who wrote plays reputedly through Shakespeare and Shaw within a few hours – an almost impossible feat difficult to explain except through some form of 'tuning in' to something like Jung's Collective Unconscious. Or does the inspiration come from dreams – day or night – which Freud interpreted through desires and needs, classifying flying dreams as creative and sexual?

Or do events like the Convair crash or the York mystery come from the flying situation that pilots call 'snowballing'? Something unexpected happens. This provides a stimulus for action to be taken. This action provides further stimulus which triggers off further action. Then that action triggers off something else happening (feathering one failed engine can lead to the loss of another). In a short time a potential disaster builds up like the plot of a novel or a piece of music, the notes being dictated by what has gone before so that a 'tune' is produced from which there is no turning back, a kind of inescapable completion of the circle on a stimulus–response basis. Once the 'snowball' has started, enormous courage is needed to stop it. Air Marshal Dowding alone stopped Churchill sending vital fighter squadrons to France after its collapse

2 VOR Omni-directional range.

as was his firm intention – a superhuman task that undoubtedly saved Britain from defeat.

In December 1944 the 'snowball' was allowed to continue. After plans had been made for British parachute troops to land behind the German lines, the presence of the 2nd SS Panzer Corps complete with heavy tanks was discovered from reconnaissance photographs. This information was known at the Headquarters of the Army Group, but was withheld from HQ XXX Corps and the First British Airborne Division and the operation was allowed to 'snowball' forward. Kept in ignorance, the leading brigade were deprived of the opportunity to increase their anti-tank capability, and the First Airborne Division advanced on a broad front – the best tactic against light opposition, the worst course otherwise – and were virtually written off.

There is just the slightest hint of 'snowballing' in that the mild XXX sent from the York merged with the SOS. But if that trap was averted, the sea over which the York was flying bordered on what was later to become infamous as the Bermuda Triangle, graveyard of ships and aircraft. Nothing had ever been found of the two Tudors that had disappeared in the vicinity. Various theories of strange magnetic and volcanic influences have been advanced as explanations of its apparent malignancy. Undoubtedly sea and earth do have their effect, but how they could influence an aircraft in this way, other than by climatic effects such as sudden and crippling headwinds – which the York certainly did not experience – has never been demonstrated.

The radio operator at Gander went on calling the York, but there was no response. US and Canadian aircraft searched the area round its last reported position for three days, but nothing was found, then or since. Only one thing was certain: the York's wing, which would have been its main support in a ditching, was above the cabin, so those on board would have been drowned within seconds of it entering the water. No body ever floated ashore, no raft was ever spotted, no piece of wreckage was ever found.

The assessors at the accident inquiry asked themselves, 'What was the cause of the accident?'

Note the singular 'cause'. That was a time when aircraft accident inquiries looked for a single cause. Yet one of the things first learned in psychology is that you can hardly ever isolate. Any event, any behaviour, any concept can rarely be taken separately. Few things have a single cause, and each cause has many facets.

For years I studied aircrew fatigue which seems a separate and easily understood concept, but like other psychologists who had worked on it, I found it a most elusive and generalized expression, introduced for human convenience rather than out of real understanding. It was not possible to isolate it from the rest of the body and its environment.

Nevertheless single causes have always been favoured by our society because multiple ones are too confusing. In spite of the fact that our minds are capable of almost limitless range and development, we live in little boxes, purposely keeping ourselves in the dark lest we be blinded and overwhelmed by all the many facets of the universe around us.

Until a few years ago, most of the findings of accident inquiries were crystallized into single causes, usually pilot error. Now fortunately multiple causes *have* begun to appear, though within limited boundaries.

What then are the *causes* that might have contributed to the York disaster? New and hitherto little known mechanical and psychological factors, certainly. But what of the other, more obscure and paranormal aspects of aircraft accidents and pressures on the outcome of events? What about imagination, fate, astrology, luck, Time, precognition, snowballing, possession or obsession, lack of communication, telepathy, coincidence – the still almost invisible stepping-stones that lead us beyond Time and Space?

There is no evidence in the disappearance of York G-AHFA to support the involvement of any of them – except one tiny glimmer. Almost certainly unknown to themselves, the captain and navigator were 'twins' in the sense that they were both born on 1 July 1922. Those who believe in astrology would say that their symbol was cancer the Crab, their sign was Water, they were ruled by the Moon and their destinies were in some way interconnected. In mid-Atlantic that night of 1 February 1953 the moon was shining from high above them on their port beam, and it was almost full.

In subsequent chapters, events will be described for which there may be many reasons and even more aspects – technical, temporal, atmospheric, personal, physiological, psychological, parapsychological. Some of the mysteries will be shown to be largely solved technically or psychologically, and yet at the same time spin off other mysteries like precognition, coincidences, fate, Time elements, other worlds – subjects of which we know almost nothing and at which we usually scoff. Radio, aviation and psychology have all suffered such treatment and have emerged into respectability. Parapsychology is still on the borderline, and the unknown beyond, in spite of being taught and preached for hundreds of years by most religions of the world, is still very much out in the cold.

But it is up this cliff face that man will have to climb to gain enlightenment. In the flying mysteries that follow can be glimpsed tiny glints of possible handholds – no more. Those handholds that we reach up to are like the handholds the first aviation pioneers reached up to in order to fly. For the air is the most mysterious, the most exciting, the most challenging of all the elements. We leave the planet, we leave the sea, we leave the earth. The air is no longer of this world. When we explore the sky, we touch infinity, eternity, time and timelessness.

And we look around and wonder. And wonder is the beginning of wisdom.

The reverse side of wonder is negativism. There are so many aspects of ourselves, our world and the universe, seen and unseen, that we do not understand. That does not mean we should close our minds to their possibilities or deny their existence.

As Koestler said, 'The limitations of our biological equipment may condemn us to the role of Peeping Toms at the keyhole of eternity. But at least let us take the stuffing out of the keyhole, which blocks even our limited view.'

2

Beyond
the Inquiry

The inquiry's view of what had happened to York G-AHFA was that the cause of the accident was 'unascertainable'.

The plug changes on number 3 engine were dismissed as immaterial. The bowser used for refuelling was checked for water and foreign matter, and none was found. It was considered that there was no reason to suppose that any mistake had been made during refuelling and rewatering at Lagens.

The question that the inquiry had set itself – 'What sort of catastrophe overtook the aircraft which had covered more than half its intended route without apparent incident or difficulty?' – remained unanswered.

G-AHFA joined *Star Tiger* and *Star Ariel* and the numerous light aircraft that had disappeared over the Atlantic after Lindbergh's successful crossing – *St Raphael*, *The Dawn*, *The American Nurse* and, strangest of all, *The Endeavour*, in which Walter Raymond Hinchliffe disappeared with the millionairess Elsie Mackay. Captain Hinchliffe was a highly respected pilot, friend of several airmen who, at the time of his death, were preparing to fly in what appeared in the late 1920s to be the answer to long-range sky travel – the world's biggest airship, R101, being built at Bedford.

To the city of Bedford, the R101 was a present-day civic pride and a future insurance against the unemployment and Depression which beset the rest of Britain. Until the advent of the airship project, Bedford had declined from her fiery and important place in history. The name meant 'fortified ford', and the Romans had recognized its importance on a curve of the wide river Ouse. The Danes had endeavoured to storm it, and after their pillaging and burning of the countryside, King Alfred had agreed with them an occupation line that ran through the heart of Bedford. Hereward the Wake and John Bunyan had been imprisoned there, the latter writing *The Pilgrim's Progress* in the County gaol at the corner of Silver Street. The city prepared for siege in the Civil War, and with the Restoration of the monarchy, her independent church was persecuted.

There was a strong Puritan tradition in Bedford and her citizens were hard-working and stoical. Cotton and lace-making, agriculture and trade flourished in the eighteenth and nineteenth centuries. But since the Great War, her importance had declined. Now the hungry thirties would be supported by the airship project. If R101 was successful, Cardington would become the world's largest airport.

And though happily the project was financed by the government, she was Bedford's ship, built by Bedford men at Bedford's suburb of Cardington. Her builders and crew lived locally. So did her designer, Colonel Richmond, whose design had cut away from the conventional Zeppelin type and incorporated his own patented gas valves, wire braces, Triplex glass windows and a lounge big enough to foxtrot all the way to India. He was a familiar figure. So was Wing Commander Colmore, Director of Airship Development, and Major Scott, Assistant Director (Flying) who lived at the Manor House at Cotton End. Captain Irwin, who commanded R101, lived more modestly in a bungalow in the Putnoe Road, while his officers and men were all admired and respected in the airship city.

Even the Minister of Civil Aviation, the dapper monocled Sir Sefton Brancker, one of the notables who would fly on her maiden voyage to India, was educated at the famous Bedford School. While the Secretary of State for Air, Lord Christopher Birdwood Thomson had taken the title 'of Cardington', when his friend, Premier Ramsay MacDonald, had created him a peer.

Bedford regarded that as the honour it was intended to be and an affirmation of his lordship's belief in the future of airships. They were pleased that Lord Thomson, being a Labour peer, showed partiality for the government R101 as opposed to the private enterprise R100, designed by Barnes Wallis and being simultaneously built at Howden.

What Bedford was less happy about was the way Lord Thomson appeared to be using the R101 to further his own political ambitions. It was rumoured that she was not yet airworthy, and that Lord Thomson was pushing designers, riggers, officers and crew beyond the limits in order to make a flamboyant flight to India and back before the Commonwealth Conference in the autumn of 1930. For the position of Viceroy of India was about to fall vacant, and Lord Thomson, the insidious rumour said again, coveted the appointment and all the pomp and splendour and truly royal powers which went with it.

In all fairness, his supporters said he deserved that position – a brave soldier, a distinguished diplomat whose persuasive powers during the last war had kept Romania from joining forces with Germany, an accomplished speaker, an intellectual, a writer himself and the friend of famous writers, a shrewd politician and the trusted confidante of his prime minister. An able man, a charming man, a determined man. A man who would let nothing stand in his way. Nothing.

There were those that spring who recalled how during his military service as a subaltern in the South African war, he had attracted the attention of Lord Kitchener. A huge jam of railway trucks was delaying the general on the approaches to Kimberely. 'Do something about it!' the furious Kitchener had ordered Second Lieutenant Thomson. To the general's delight, young Thomson had rounded up gangs of men and ordered them to overturn all the trucks and allow the general through. Thus Thomson's upward climb had begun.

But of recent years Thomson's pursuit of power was not wholly for its own sake. During his appointment as military attaché to Romania, he had met and fallen in love with a princess connected with the royal family, and it was said he wished to lay the Indian vice-regal crown at her feet.

'The flight to India must take place,' Lord Thomson had said as far back as the previous November. If not by Christmas, then soon after. But Christmas had come and gone, and now snags kept being discovered in R101. She rolled. The valves of her huge gas-bags chattered, letting out the gas, there was chafing of the girders against the goldbeater skin of the bags. Worse still, it was discovered, she was too heavy. The empty ship weighed 113.6 tons and instead of the hoped-for ninety tons of fuel and payload, she could only lift an inadequate thirty-five.

She was returned to her huge steel hangar to be pared down. Tanks, cabins, toilets, a look-out at the top of the ship, servo-motor steering gear were removed. The Triplex glass in her windows was replaced by a type of light celluloid called Cellon. Lord Thomson was assured that the airship would be ready for her trip to India by the second week in March.

But the 1930 Ides of March came and went and the airship was not ready. And though R101 could now get to India in her lightened state, she would not be able to get home, because of hydrogen loss in the heat of Karachi. It was tentatively suggested that she be cut in half, and another gas-bag inserted to give her the required lift. Perhaps because the ship was already assuming a life and personality of its own, some of the technical advisers opposed this surgery. But Richmond, her designer, nick-named Dopey, for he began his career as a chemist, found it feasible. It was decided that the operation would take place early in July. A bay forty-four feet long containing another gas-bag would be inserted in her middle.

Meanwhile on 23 June while she still awaited that major surgery, a lightened R101 was walked out of the hangar. This was always a tricky operation but a popular one and Bedford rejoiced. Crowds gathered along the roads near the huge gaunt hangars which were higher than Nelson's Column and bigger than Westminster Abbey. The approach roads were jammed with vehicles. Ice-cream carts, chestnut roasters and postcard sellers gave the place a holiday atmosphere and, more important still, hundreds of the unemployed were taken on to help, to hold the mooring ropes and the handling bars.

The big hangar doors were rolled back. Standing in the control car, R101's captain, Herbert Carmichael Irwin, gave the order 'Walk!'

Between the hangar doors with only inches on either side to spare, the great dirigible was led out by miniscule men. So finely balanced was she that her 113 tons floated like a dandelion puff. Then suddenly a playful wind swept her sideways, and rent a tear 140 feet long in her outer cover. That was repaired but the same thing happened the next day. Time was important. This tear also was quickly stitched and treated with a rubber solution. Too quickly, in the opinion of many. Nevil Shute, the famous author, who was working then as a designer on the rival R100, was over at Cardington on a visit and said, 'The effect was to make the cover so flaky, you could put your finger through it.'

His disquiet was shared by Captain Irwin. A calm, quiet and brave man, he was increasingly concerned about the airworthiness of his ship. He had seen the reports made by a very conscientious aircraft inspector called McWade. Mysteriously, some of these reports had not been forwarded to the Air Council.

The holder of a short service commission in the RAF and an Olympic athlete, Irwin was not given to flights of fancy or fear in any form. Yet he was Irish and his wife was Scots, and they shared the Celtic quality of insight. Something was wrong with the ship of which he was so proud, something that must be righted before they left for India.

Then on 28 June Irwin flew R101 to the RAF Hendon Air Display. It was to be her last flight before she went into the hangar again for her surgery. He found the airship 'heavy' – that is, she would have slowly descended if up-elevator had not been applied continually. She responded sluggishly to the controls. The valves chattered, letting gas out every time.

Flying over the Royal Box the previous day in rehearsal, R101 dipped her nose in a salute that was as low as a curtsey, and was only with difficulty pulled out by the sweating elevator coxswain. Captain Irwin filed his report.

The next day R101 was taken back to the hangar. She was cut in half. Work began on the new bay which would bring her length to 777 feet. At the same time, her sixteen other gas-bags were examined. These were made of the intestines of over a million oxen, the best material to prevent seepage of the $5\frac{1}{2}$ million cubic feet of gas carried. It was found that all but one were riddled with holes – number 11 had 103 – caused by rubbing against projecting parts of the airship frame.

Such chafing had been the subject of several of Captain Irwin's reports, but Colonel Richmond and the Director of Airship Development, Wing Commander Colmore, had only been able to suggest padding the offending parts. Four thousand pads had been made for this purpose, but some of the chafing continued. And once more the inspector, Mr McWade, sounded a warning.

Already there was tension between him and Richmond. McWade was tired of having his technical opinion either overruled or ignored. So instead of reporting his views to his immediate superiors (Richmond and Colmore) he went over their heads and sent a confidential letter direct to Air Ministry in the hope that it would be read by Air Vice-Marshal Higgins, the Air Member for Supply and Research, or by Lord Thomson, the Secretary of State himself. After explaining his objections McWade wrote, 'until this matter is seriously taken in hand and remedied, I cannot permit to you the extension of the present "permit to fly" or the issue of any further permit or certificate.'

They were strong words which might have been effective. But they were never seen by Air Vice-Marshal Higgins or by Lord Thomson. Instead, the letter was sent back to Cardington for Colmore's attention. McWade was ordered peremptorily to get on and pad.

During the three months which it took to insert the new bay and pad the girders, the crew had little to do. Some of them – Sky Hunt, the chief coxswain, included – were uneasy about what they called 'her stomach operation'. Their wives worried. Living in a close-knit community at Shortstown just by the entrance to the field, they endeavoured to scotch their own misgivings. Then on 29 July the private enterprise R100, under the command of Captain Ralph Booth, a close friend of Irwin's, made a successful crossing of the Atlantic to Montreal. On board with him was Major Scott, the Assistant Director of Airship Development.

A fortnight later she flew back. Private enterprise had won. By the double flight to Montreal, R100 had initiated the linking of the Empire by air, the All-Red-Route so beloved of Lord Thomson. It was more than ever imperative that R101 should show her paces to India and back. Besides, the Labour government was tottering. Lord Thomson might soon lose his exalted position. Yet all would not be lost if he could fly in R101 to India before the Commonwealth Conference and be seen as the obvious viceroy.

Meanwhile the forcibly idle crew foregathered at the Bell Inn at Cotton End, known locally as Betsy's, and run by a popular warm-hearted landlady called Betsy Bunker. The officers usually favoured the lounge bar of the Bridge Hotel, overlooking the river on a site burned by the Danes, for, as a chronicler of the time wrote, 'the Danes ever burned as they went'.

At both places, the talk that summer was of airships. The ground crew were of the opinion that Captain Booth of the R100 should have received a knighthood. So should Major Scott, their assistant director who had helped him fly her, and who had made a double crossing to New York ten years previously in the R34. Alcock and Brown, who had crossed the Atlantic only one way and then crashed, had been knighted. Why not Scott and Booth for a greater feat?

Scottie, a stout, bluff pipe-smoking extrovert, was especially popular with the ground crew. Captain Irwin they respected and trusted, but he was a quiet man, difficult to know and he rarely unbent. And though he was known as 'Lofty' affectionately by the ground crew, it was never to his face. Nor did anyone but his wife and his closest friends use his special nickname 'Bird'. A name which symbolized his passionate love of the air, his belief in its conquest, and the freedom that he found only there. What it could not symbolize was his deep sense of responsibility as captain of R101.

He had used his enforced idleness wisely. He had worked on further series of tests for R101 when she emerged in her new elongated form. He had resolved he would not take her to India till these tests were completed. He trusted his crew, but he had mixed feelings about the fact that Scottie would be accompanying them. Scottie was ambitious. If the flight was successful, Scottie would get the knighthood he deserved. But Irwin's friend Ralph Booth, R100's captain, had told him Scottie had been a bit of a liability on the epic Montreal trip. He had steered R100 smack into a thundercloud which had tossed the airship up 4000 feet and rent her outer cover. At the welcoming receptions there had been embarrassing arguments about precedence. But his experience would be invaluable; as would Squadron Leader Johnston's, also a veteran of that flight and the finest navigator of them all, but inclined to be quick-tempered.

The atmosphere while the airship crews waited for further action became strained and difficult. In the playgrounds of the schools, there was fighting between the children of the R100 and R101 crews. On 16 August 1930 after her successful flight to Canada and while the R101 was still being lengthened, R100 was brought to Cardington where Lord Thomson greeted the crew amidst much congratulatory conviviality. Some of the R101's crew were detailed to refuel the R100, after which three fuel tanks failed and broke through the outer cover.

According to an engineer on the R101 reserve crew, though Atherstone did not mention it in his diary entry of that day, an incident then took place.

The men had been 'feeling no pain' and had overfilled the tanks, petrol seeping through to form a highly inflammable puddle underneath the airship. One of the culprits made a sassy remark to Johnston, who clipped him on the jaw. For an officer to strike another rank was a court-martial offence. But there were extenuating circumstances beyond Johnston's justifiable indignation. For the past two years, he had been receiving warnings that the R101 would go up like a bomb.

In the clear light of day, these warnings to a brave man like Johnston were plainly nonsense. But sometimes when he was keyed up, he remembered them together with the fact that they came from two people whom he greatly respected. The first was Emilie Hinchliffe, the widow of the famous aviator and his old friend, Captain Hinchliffe who had disappeared over the Atlantic with the shipping heiress, the Honourable Elsie Mackay. The other was no less a person than the famous author Sir Arthur Conan Doyle.

'The R101 is doomed!' Emilie had told him. 'I beg you not to go!' She had been to see Johnston at Cardington, telling him that while in communication with her dead husband through a medium called Eileen Garrett, Hinchliffe had specifically told her to warn his friend Johnston of the hazards of R101. Dear devoted Emilie had done exactly as the spirit of her dead husband had told her. And when Johnston had simply laughed at her warnings, she had come again bringing with her Sir Arthur Conan Doyle. He praised this Mrs Garrett for her absolute integrity, and for his part begged Johnston to take heed. Not only had Captain Hinchliffe spoken through Mrs Garrett, but the medium herself had seen strange daylight visions of an airship flying over London, always with the knowledge that it was the R101, not the R100 or any other. And as she watched, suddenly the vision would burst into flames and disappear.

Strangely enough, while Johnston roundly dismissed these warnings of disaster and fire, the talk at Betsy's tended to hark back to an incident when R101 was first walked out of the hangar on 12 October 1929. Captain Irwin was in the control cabin, a gondola immediately below the belly. On the ground, her designer Richmond and the Director of Airship Development, Colmore, watched the operation with held breath. Of all the watchers, and there were thousands that frosty morning, only they already knew she was too heavy to fly to India and back.

The ground handlers reckoned she was weighted by the five massive Beardmore Tornado diesel engines, built primarily for Canadian railways and slung in pods underneath her. But Bell and Binks and King and Gent, all R101 engineers and frequenters of Betsy's, were proud of them. They dismissed as irrelevant the fact that Scottie, now watching the walking out from just beside the mooring mast, had married a Beardmore. They pointed out that diesel oil was less inflammable than petrol, and fire was the dread of airshipmen.

The Beardmore engines were not started that birthday morning as Irwin eased her gently through the high metal doors and out into the chill. Dawn was breaking and they could see the outlines of the charabancs that had brought the sightseers, and the dense crowds thronging the perimeter roads. A great cheer went up from thousands of throats as the great tail cleared the hanger, and the walkers began to lead her across the grass, a captive whale led by ants. It was only afterwards that there were questions as to why R101 had never been named or christened.

To leeward of the mast, Irwin gave the order to pay out the main haul cable. Once connected, he ordered the side guys to be run to the ground position. As she lay tethered and still, Scottie, relaxed and ebullient, raised a cheer from the crowds by climbing like a monkey up into the control car from the ground. Happy as a schoolboy, he had insisted on taking over the controls for the tricky locking-on manoeuvre when the egg-shaped nose was eased into the cup of the mast.

He nodded at Chief Coxswain Hunt. He put the megaphone to his lips and ordered the working party, 'Let go!'

Up R101 rose. Then, meeting the icy morning air above, it bumped against it like a toy balloon against a ceiling.

The order was given, 'Release ballast!'

Down came a great shower of water that sent the ground crew running for shelter. Then suddenly they were all running after something else. Something streaking across the grass ahead of the nose of the ship.

A small brown furry animal – a hare. Which then, as with a single purpose, three hundred men had begun chasing off the field as if their lives depended on it.

Or as if they remembered the age-old superstition, that a hare running ahead was a warning of death by fire.

Death by fire had been foretold for one of the most distinguished passengers for R101's maiden flight to India. Sir Sefton Brancker, Minister of Civil Aviation, a fearless airman and an incorrigible ladies' man, had formed an attachment for Miss Auriol Lee, a famous actress-producer. And she, who returned his admiration and affection, had in a moment of anxiety asked an astrologer to cast his horoscope. Back had come the prediction that he would meet his death by fire. Oddly, this underlined something a clairvoyant had told him in Paris some years before, that after 1930 she saw no future for him. Nothing.

Stranger still, at a London party in mid September he had been introduced to a strikingly handsome woman who, brushing aside his gallant courtesies, had come straight out with the words, 'R101 will crash in flames.' Her name, she told him, was Eileen Garrett.

But it was none of these gloomy predictions which made Sir Sefton take issue with Lord Thomson that day at the end of September when, her modifications completed, the elongated R101 had been led out of the hangar. He had heard of the tests which Captain Irwin had requested. He knew that Colmore wanted more mooring masts placed at intervals on the route to India. And yet he had heard that now Lord Thomson was pressing Cardington to give him an early departure date.

'Colmore and Richmond assure me I can get to India and back before the Conference. The crew are quite happy to take her,' Thomson told him brusquely.

But there he was wrong. Captain Irwin was becoming increasingly worried as on the one side pressure from Air Ministry to go increased, and on the other bad weather prevented even beginning the trials he had asked for. Gradually he was becoming certain that it would be irresponsible to take an inadequately tested R101 on a new route full of climatic changes and unknown weather hazards. The safety of all on board was ultimately in his hands.

Then on 30 September the weather improved. Air Vice-Marshal Dowding, who had succeeded Air Vice-Marshal Higgins and who was ultimately responsible for the airship programme, was informed that a 24-hour trial flight was about to take place. Much depended on whether he would be able to come. Regretfully he had to decline because the flight was so long that it would coincide with a previous engagement. But on 1 October Wing Commander Colmore found that the test could be curtailed, so that Dowding would be back in time for his appointment.

That day the Air Vice-Marshal flew up to Cardington. R101 was already locked on to the mast and the test began. It was the first time Dowding had ever been up in an airship.

The weather was, in Richmond's words, 'very perfect'. A few hours after they were airborne the starboard forward reversing engine began to leak. At night, while Dowding, kept in ignorance of any trouble, slept in one of the bunks, a new washer was fitted, but as a safety precaution, the engine was stopped and the port engine was reduced to half speed.

After less than seventeen hours trial, R101 returned to the mast. Two attempts were necessary to moor her. Afterwards Irwin filed his usual meticulous report, a carbon copy of which he kept in his desk. Colmore and Richmond expressed themselves as delighted with the extra lift from her new gas-bag. Scottie seemed satisfied. A gassing and mooring party had already been sent to Karachi to await R101's arrival.

Tired though he was after the flight, Colmore went up to Air Ministry that Thursday 2 October 1930. Lord Thomson was anxious to hear at first hand how the test had gone. Before accompanying him into the Secretary of State's office, Air Vice-Marshal Dowding had said, as if mindful both of his own ignorance and the undue haste, 'You are my adviser and whatever line you take with the Secretary of State I shall back you up.'

But Colmore's line was optimistic. The only question was *when*. Tomorrow or the day after? Saturday evening was chosen to allow the crew to rest and check the ship and for R101 to cross France by night.

Air Vice-Marshal Dowding raised the point that a full power test had never been carried out, and suggested this could be done as soon as R101 slipped the mast for India – a suggestion which Colmore did not gainsay, though it was scarcely feasible, fully laden and committed as R101 would be. It was also a suggestion which was not put in writing. What was put in writing was a letter from the aerodynamic expert, Professor Bairstow, to Air Ministry, telling why his report on the elongated R101 was not ready. 'The conditions between R101 now submitted and those of the original design on which our original report was based surprised us by their magnitude . . . we have not had time since receiving essential information from the Royal Airship Works to prepare a sufficiently considered written report.'

Nevertheless the certificate of airworthiness was prepared. Preparations went rapidly ahead. So sure was Captain Irwin that the decision was wrong that he decided on the only possible option remaining to him. He would refuse to take R101 on Saturday to India. He had a short service commission in the RAF and this he would have to resign. Without him the flight would surely have to be postponed.

But he discovered there would be no question of the flight being cancelled. Too much was at stake. Colmore had been told that without the trip to India, there would be no more money for airships. If Irwin refused, his friend Squadron Leader Ralph Booth would be ordered to take his place. And as a regular officer, Booth could not refuse. So Irwin either had to take his ship into danger or send his friend.

It was with a heavy sense of responsibility that Irwin checked the ship on 3 October – a sense shared by many of the officers and men. Irwin's first officer, Lieutenant-Commander Atherstone, who had been persuaded to come back to England from his farm in Australia to lend his expertise to the airship project, wrote in his diary, 'we all feel the future of airships very largely depends on the show we put up'. He mentioned unknown factors and hoped for good luck.

Meanwhile that Friday it had come to Lord Thomson's ears that Irwin had misgivings. He telephoned the captain direct and accused him of being an obstructionist.

He used the same technique on Sir Sefton Brancker later that day. The Minister of Civil Aviation had the effrontery to come into his office, and tell him point-blank that Colmore wanted more mooring masts, that the R101 was not ready, and that the flight should be postponed. Worse still, he had suggested that with the Commonwealth Conference coming up, it was better to leave R101 at her mooring mast, let the delegates see her, admire her, dine aboard her.

Thomson's retort was, 'Of course if you're afraid you need not come along.'

'I got rapped over the knuckles and I got no change,' Brancker was to confide the next day to his old friend and ex-comrade, Major Villiers, now an intelligence officer at Air Ministry. Major Villiers had undertaken to drive Sir Sefton and the well-known pilot Winifred Spooner from Henlow to Cardington for R101's departure.

It was a gloomy ride. 'No change at all,' Brancker repeated as, miles ahead over the countryside, they glimpsed the silver shape of the airship, riding at her mast like a basking fish.

That fish shape, though closer at hand and many times bigger, had been clearly visible through the kitchen window of the Irwin's bungalow, Long Acre. Olive Irwin purposely avoided looking at it. 4 October, the day she had so much dreaded, had come. With characteristic thoughtfulness her husband had arranged that her sister and brother-in-law should stay with her and, unbeknown to her, had talked long into the night with his brother-in-law about his conversation with Lord Thomson and his own reservations about the airworthiness of the ship.

But in the morning, when he donned the specially designed R101 uniform of navy blue reefer jacket with gold bars on the epaulettes and peaked blue cap with the insignia R101 picked out in gold thread, he seemed cheerful and as always calm. Victor Goddard, who had been considered for command of R101 and who was with Irwin flying airships in the Great War, described him as 'the gayest of all'. He was well able to assume a confident poise, despite the fact that a new problem had reared its head.

Or rather an old one had not been satisfactorily resolved – the question of the captaincy of R101. Officially the position was his. Yet ironically he was one of the most junior officers aboard her. Scottie was his senior in rank, senior in experience and senior in the airship hierarchy. Any mariner or airman would recognize that there is nothing potentially more dangerous than divided command.

Scotttie had already made clear his dissatisfaction with the R100 flight to Montreal, when he had felt himself placed in an inferior position to Squadron Leader Booth. Both Irwin and Scottie had separately seen the administrative officer at Cardington to clarify the captaincy position for *this* coming trip. The administrators tended to play it both ways. Of course Flight Lieutenant Irwin was the captain. But what of Scottie? Official-in-charge-of-the-flight was too vague for him and not good enough. Passenger was worse.

He made his complaint vociferously. The administrators and the press officer and Wing Commander Colmore agreed a wily compromise. The first communiqué to be sent out to the world when R101 slipped the mast was drafted while Irwin and his crew were on board making their final checks. It read, '*Airship R101 left the mooring tower at Cardington at . . . . . hours GMT, 4th October on the first stage of her flight to India. The flight is being carried out under the direction of Major G.H. Scott, CBE, AFC, Assistant Director in Charge of Airship Flying.*' Previously a passenger, Scott appeared the next day in uniform.

Irwin's name as captain was lost in the welter of names that followed.

Scott read it and told the press officer, 'it was entirely satisfactory and accurately represented his own view of his position'. But there was no time, as preparations for departure were proceeding, for the press officer to speak to Irwin on *his* position in relation to Major Scott.

The captaincy matter had not really been settled, and the morning weather report indicated a depression to the north-east, though Giblett, the meteorological officer who was going with them, had indicated it was moving away.

That same morning there was a merry party at the Bridge Hotel, Bedford, attended by Johnston, Colmore, Richmond and other officers, together with a host of well-wishers. The landlord presented them with a nine-gallon cask of beer, its rims edged with silver, R101 engraved in silver on its side above the crossed flags of Britain and India.

At Betsy's, there were few customers. Engineers like Binks and Bell and young King, riggers like Church and Radcliffe, electricians like Disley, were already up at the field, the meagre belongings they were allowed to take already packed and being weighed.

By lunchtime at Cardington the whole station was palpitating with activity. Men were shouting through megaphones. Gouts of steam puffed out from the donkey engine that worked the lift. There was the padding of rubber-soled feet as the crew ran up and down the 197 steps. At the masthead, there was a hissing as the gas topping-up continued. From inside the ship, as she flew at the masthead, could be heard the clanging of telegraph bells.

At 15.03 another weather report was issued . . . '*the occluded front over*

France this morning has now passed eastward while a trough of low pressure off western Ireland is spreading quickly. Cloud is increasing to ten tenths and falling to 1000 feet. Rain will spread from the west, probably reaching Cardington tonight.'

The weather was closing in. A storm of wind and rain was reaching out towards them. Irwin must have paused at that point and considered calling a delay. But already Scottie was rushing around like a bull in a china shop shouting, 'For God's sake, let's get off! Where the hell are the passengers?'

They came at a leisurely pace. The Air Ministry had forbidden all flights of aircraft in the vicinity of Cardington, so most of the passengers arrived by car or train: Bishop and Bushfield from the Air Ministry Inspectorate; Squadron Leader Palstra of the Royal Australian Air Force; Squadron Leader O'Neill, recently appointed Deputy Director, Civil Aviation, India. Of the many who embarked on R101 that day, his feelings were probably the most divided. On the one hand he was worried about his wife, who had returned to England weeks ago for an operation by the famous gynaecologist, Sir Henry Simson. Naturally Squadron Leader O'Neill had taken leave to be with her. But because Simson was also gynaecologist to the Duchess of York, Mrs O'Neill's operation was postponed until the Duchess was delivered of Princess Margaret Rose.

Torn between missing his return passage to India, and leaving his wife before her operation took place, he had had the fortune – good as he had interpreted, tragic as it turned out to be – to meet Sir Sefton Brancker who had waved his magic wand and secured him a berth aboard R101. By 4 October Sir Henry Simson had successfully operated on his wife. Sad though he was to leave her, O'Neill was full of anticipation for the inaugural flight.

So apparently was Lord Thomson, still taking tea with his valet at the village hotel at Shefford, while the other passengers, Sir Sefton Brancker included, were being weighed and searched for inflammable material.

'I won't say goodbye,' Sir Sefton said to the Villiers, as he took Winifred Spooner aboard to admire the luxurious lounge and dining-room, the metal-lined smoking-room, the fitted carpets in Cambridge blue. 'I'll come down later to say farewell properly.'

Winifred Spooner had come back and Villiers was still waiting for that farewell when Lord Thomson arrived. The cases of champagne going on board were as nothing to his personal baggage. Peering out of the bowels of the ship, the crew, who were only allowed to take a change of uniform and underwear, a pith helmet for India, tooth and shaving brush, muttered angrily as besides nine pieces of luggage, there was unpacked a huge roll of carpet.

'To impress the Egyptian king,' he had told Buck, his valet. 'His Majesty will dine aboard R101 when we reach Ismailia.'

That carpet, added to the silver and napiery, the specially fired Wedgwood dinner service, would add unconscionably to the weight.

'Not to mention the weight of him and his blinking valet,' Bell murmured, watching through the slit of the aft engine cupola. But neither were weighed. Nor were they searched. Had they been, the searchers would have found an unexpected article. Treasured deep in his most personal and stoutest tin trunk was a silver slipper, once worn by his beloved Princess Bibesco. It was his talisman, his good luck omen, his token for the future. Ironically, it was one of the few objects which survived the holocaust.

'And shall you be making the entire return journey to India, sir?' asked the clamouring reporters who clustered around him, their camera flash-bulbs popping. Everywhere, lights were flashing like a fairground – car headlamps, rotating warning lights, searchlights – reflected now on the lowering clouds.

'Why not?' Lord Thomson smiled, 'I am under orders to be back in London by the 20th, and I don't expect to have to change my plans.'

He had used similar words at Number Ten the day before to a far more important person, his Prime Minister, Ramsay MacDonald. Moved as he sometimes was to a mood of sudden melancholy, Ramsay MacDonald had put his hand on Thomson's shoulder and begged him to change his plans for the India trip. His eloquent eyes had taken on that luminous inward look of the seer which some people believed him to be. Shaken by his vehemence, Thomson had asked, 'Are you ordering me not to go, sir?'

'No. I am *asking* you, not ordering.'

'In that case . . .' Thomson had smiled and given his confident reply.

But for all his confidence, he had made a hasty will, had insured his own life and that of his valet, and had paid a quick last-minute visit to his aged mother in Devon.

Now bold and smiling, watched by a hundred and fifty thousand people, he stepped on board. A moment afterwards a messenger came running up waving a piece of paper – a last minute reprieve, some of the waiting wives thought. A hoped-for postponement, the Villiers prayed, still waiting to say goodbye to Brancker. But it was the belated certificate of airworthiness, which had been issued but not delivered.

A few hundred yards away, the man responsible for the certificate, Air Vice-Marshal Dowding, watched from the other side of the tower as the engines were started. All except the forward starboard one fired. Then in a shower of back-firing sparks, it started.

At long last, all five engines turning, R101 slipped from the masthead – and immediately lurched towards the ground. In the rush the heavy roll of carpet and extra cases of champagne had been stowed in the nose and forgotten.

Ballast was released. Seven times a cascade of water descended on the hushed crowd below before the bows rose. To cheers and whistles, R101 rose higher and began a wide circuit of her native Bedford. The crew, in their new brown boiler suits and soft shoes, ran to the windows to flash goodbye to their families with the shiny metal pen-type torches with which they had been issued.

Then the torchlights faded from view, as R101 set course into the rain-filled darkness for her passage to India.

One of the mysteries of the R101 is why she went at all.

With that meteorological forecast, there must have been a conference. It would be in Scott's nature to take a chance. Irwin would perhaps have bided his time. He would not have wished to antagonize Scott. He might simply have been overruled, and Scott would have taken over. He must have comforted himself that in an emergency they could come down on to the ground at Paris, where an emergency handling party of 300 men had been arranged. Alternatively, if things got too bad, despite Lord Thomson's presence, they could return.

Almost immediately, the flight was in trouble, in direct contrast to the champagne and bonhomie in the passenger lounge. Torrential rain thundered on to the envelope. The oil pressure in the aft car began falling. Bell shut the engine down. Then, deciding it was only the gauge, he started it up again just as his mate Binks climbed down the ladder into the cupola for the eight o'clock watch.

Watched by thousands, R101 droned over New Barnet as the clock was striking nine. She was so low that numerous people looking up thought she was going to crash. But Scott's wireless message over London was '*All well*'.

Then the oil pressure in the aft car began to fall again. Engineer Officer Leech, with his top two engineers, squeezed with Binks into the tiny cupola to try to find what was wrong.

Very slowly, into the teeth of a gathering head wind, R101 continued to the coast. Over Fairlight she was reported below the level of the cliffs, less than fifty feet above two houses, one eye-witness said, and crabbing. A Hastings fisherman nearly telephoned the lifeboat.

Back went another message to Cardington: '*Crow* [=four engines running] *at 21.35 GMT crossing coast in vicinity of Hastings. It is raining hard and there is a strong south-westerly wind.*'

In the aft car, cramped together in the tiny metal cell, the four engineers were still struggling to get the engine repaired. In the port mid-ships car Cook could see the long tongues of spume less than the airship's length below him.

Half way across the channel, they were so low that Atherstone, then officer of the watch, grabbed the elevator wheel from the height coxswain and began winding it back, saying, 'Don't let her go below 1000 feet!'

At 23.00 hours the watch changed again. The engineers and riggers gulped down cocoa and sandwiches in the rest-room and slithered along the oily companionway to their positions. Bell relieved Binks in the still silent aft car where Leech and the others were still working on the engine.

And then suddenly things improved. It stopped raining. The wind dropped. The chattering of the gas valves ceased. A flashing white beam announced the lighthouse on the French coast near St Quentin. And at last the top engineers got the aft engine started, and climbed thankfully back inside the envelope.

It was only a brief lull. No sooner was R101 over France than the wind came howling back. She was still flying 'heavy'. At Poix airfield the manager reported her at only 350 feet. Though her airspeed was 60 mph, her ground speed was down to 31. And she had another 2300 miles dog-legging round the Alps before she reached Ismailia, Egypt, the only other mooring tower before Karachi. The new forecast showed worsening weather.

Out to the world went another of Scott's reassuring messages: '*After an excellent supper, our distinguished passengers smoked a final cigar and having sighted the French coast have now gone to bed to rest after the excitement of their leave-taking. All essential services are functioning satisfactorily. The crew have settled down to watch-keeping routines.*'

At this point, there must have been some exchange between Irwin and Scott about going back. No British airship had flown in such weather over land. The first revised communiqué (which R101 would have received) had been transmitted to the world. Perhaps now Irwin saw it for the first time, and realized its implication that Scott would take *all* the decisions.

Just before turning in, Brancker went to the wireless cabin to send a message to his friend Auriol Lee at the Elysée Hotel in New York. He found Squadron Leader O'Neill who had just sent a message to his wife in the Hammersmith Hospital. Having sent 'Off at last. Blessings. B,' Brancker stayed to chat with Johnston who was navigating. He must have read the latest weather forecast, looked at Giblett's weather map, and seen the projected route that would take them over the roofs of Beauvais and Beauvais ridge.

He would have had to have been a good deal less intelligent than he was not to have been reminded of the time two and a half years ago when he nearly lost his life flying over Beauvais. Then he was in an Argosy which was suddenly caught and sucked down in a sort of skyquake. Passengers were hurled to the ceiling. Two of them were injured. The door was torn off and there was considerable damage.

On questioning the Air Ministry meteorologist about the cause of that phenomenon he had received this reply:

Severe local squalls . . . gave rise to considerable vertical disturbance in their neighbourhood. It seems possible that Argosy was caught in the rear of one of these squalls where there would be a decided downward current. If as seems likely, the machine was at the time on the leeward side of Beauvais ridge, the downward current would have been intensified.

Had it been a warning, like the other warnings he had received?

Brancker must have recognized the similarity of the weather then with the weather now. It must have entered his head that forces were gathering to bring about almost a carbon copy of that near catastrophe of two years ago.

Except that Argosy was a manoeuvrable well-powered aeroplane, and R101 was a huge dirigible 777 feet long.

After his labours on the aft engine, Engineer Officer Harry Leech went into the smoke room for a well-earned cigarette and a whisky. He was still there at 02.00 GMT, when the watch again changed.

By an odd coincidence, it was the moment that the clocks in Britain had to be put back from British Summer Time 03.00 to 02.00. By a further coincidence, at that precise moment the wind vane at the Beauvais weather station swung thirty degrees right to 240°.

The clock was just striking two in the town of Beauvais when the noise of engines woke Monsieur Bard and his wife. Getting out of bed, they saw R101's lights low over the roof beyond the tennis courts. Further down the street, Monsieur Patron also heard them and went outside and saw one white nose light and one green starboard light. Out for a night's poaching, Monsieur Rabouille, an employee at the local button factory, was just setting up his rabbit traps in the Bois des Coutumes when he heard the engines rumbling over him.

'There was,' he later told the inquiry, 'a tempest from the west.'

Louis Tillier, a shepherd in the clover field at Bongenoult, heard them too. The door of his hut was half open. On looking towards the woods, through heavy rain he saw the airship lights in a row, like the illuminated windows of a passing train.

Precisely at that time, behind those illuminated windows, the second watch were getting out of the bunks in the crew quarters, grabbing a cup of tea provided by 'Carbolic' (so-called because his tea always tasted of it) Megginson, the eighteen-year-old cabin-boy chosen from 500 applicants for the job, and racing along the companionway, the riggers to their stations, the engineers to climb out into the stormy air down the ladders to the cupolas, while the relieved first watch climbed gratefully into the still-warm blankets.

All except Binks, the second engineer of the aft engine, who was late. It was 2.05 before he clambered into the hot noisy cupola, and Bell, his first watch mate, was just giving him a piece of his mind when suddenly he was flung forward against the Beardmore.

At the same time in the smoking room, Engineer Officer Leech was just reaching for the soda siphon when it and the glasses slid down the table on to the floor. It was just the nose dipping, he thought, as they came back to the horizontal. He shrugged his shoulders, picked up the glass and the siphon and put them back on the table. A few seconds later, they slid off the table again and he was sent crashing against the wall. Above the noise of the engines, he suddenly heard running footsteps and 'Sky' Hunt's stentorian voice, 'We're down, lads!'

Up at the nose, rigger Church was just going back to the crew quarters after his watch when he received a message from Irwin to release the half-ton water ballast forward. He was just going back as the airship dived that second time. Panting and sweating, he was crawling towards the tank when he felt under him a slight impact. The airship shuddered, gave a little jump, then crunched softly back to earth. Church heard the cracking of girders like fir trees falling. Then there was a *whoof* like petrol spilling on a concrete floor and he was enveloped in a white sheet of flame.

Badly burned, Church struggled free. Leech managed to fight his way out of the smoking room. So did Savory from the starboard midship car, Cook from the port. Rigger Radcliffe and Disley, the electrician, escaped from the wreck but not without injury.

But in the aft car, Joe Binks and Arthur Bell were trapped. The engine had stopped. They were still bumping along the ground as R101 telescoped. Smoke filled the cupola. The light had gone out. But they needed none. The interior was lit by the sudden inferno outside. Flames were licking under the door, towards the petrol tank for the starter engine.

'If that goes up we're done for, Joe.'

Suddenly they both realized they had only smoke to breathe. No air.

'We're done for anyway. We're suffocating.'

'Better than being roasted, Arthur.'

Bell nodded.

'This is our lot,' Binks said. Sweat ran down into their eyes. They knew if they opened the door they would explode.

Nothing could save them. With dignity, they shook hands. Binks thought of his parents and his friends at Betsy's bar. Then summoning his faith, he thought of his Saviour and lifted up his face. Suddenly he felt something wet on his scorched cheeks. Just a few drops at first. Then a whole cascade, as water poured over them, deluging the flames so they could get to the door and make their escape.

A ballast bag had burst in the disintegrating carcass above them, which explained the miracle. They were saved – as were six others – by a British racehorse owner called George Darling, who happened to be staying with a French friend nearby and came rushing to the inferno, together with the people from Beauvais and the 51st French Infantry stationed in the town.

The eight survivors were taken to hospital. But the forty-six others had died in the fire. Irwin was last seen calmly giving orders in the half-crushed control cabin surrounded by flames. Lord Thomson, Colmore, the Director of Airship Development, Richmond, R101's designer, Scott, O'Neill, Atherstone, Palstra, Hunt and the other crewmen – all perished.

One of the first things found as the flames died down was Brancker's monocle – blackened but unbroken. The watch given to engineer King for his services as a member of the R33 breakaway crew was still going (and was later exhibited by its makers in their shop window). The stamps he carried in the back were still there, unmarked. Most of the other watches were stopped at 02.09. The silver slipper belonging to Princess Bibesco was found intact and unscorched in Lord Thomson's baggage.

Disley, the electrician, managed to telephone the Air Ministry with the news. The Prime Minister was informed. The French wireless broadcast the message '*G-FAAW a pris feu*' all over the world. Irwin's friend Booth was told and went immediately with his wife to the bungalow in Putnoe Road to break the news as gently as he could to Olive Irwin, trying first to wake her sister.

But Mrs Irwin was already up. 'You don't need to tell me,' she said, 'I already know.'

Many wives had premonitions that night. Atherstone's dog howled throughout the early hours of the morning. And there were stories of a curious phenomenon at Cardington. At 2 a.m. one of the telephone flaps came down indicating that someone wished to make a call from one of the offices. The man on watch at the switchboard saw it was from Captain Irwin's office. Knowing he was not there, he flicked it back up. Seconds later, down it came again. Fearing burglars, he had the office searched. But it was quite empty.

Dawn over Beauvais illuminated R101 lying on the brow of the ridge – her nose in a little wood, still fitfully burning. All night, nuns from the Convent at Beauvais had kept vigil over the dead. Villagers had brought flowers and filled the lorries that eventually conveyed the remains to Beauvais. Smoke rose steadily from the skeleton, and pieces of gilded promenade deck, steps from the staircase, chain harness from the seventeen gas-bags, the big round valves that 'chattered' so often hung from a spider's web of broken girders, while pieces of blue and gold crockery – Bedford's gift – littered the burned grass.

But the RAF flag was still flying from her tail, the blue cotton stiff in the breeze, when the Air Ministry investigating team, including Ralph Booth, arrived on the site that Sunday afternoon.

Back at Cardington, the wives of the living crewmen tried to comfort those of the dead. Through the streets of Bedford marched the Salvation Army band playing the Dead March from Saul.

That afternoon, during another seance with Eileen Garrett, Mrs Hinchliffe's husband again came through.

'That damned job could not stand up under air pressure and currents, and was caught and had to battle for two hours against the elements. If I, not heavily loaded, and with cubic inches to spare, could not stand up against what I met, what the devil could they expect, having had the knowledge that the weather had broken and that there were gales ahead? In the event of a storm, she was too top-heavy . . . she wasn't able to rise.'

Bearing in mind that the seance was only around twelve hours after the accident and that no real information had been published, much of it is remarkably accurate.

'Hinchliffe' finished up with a furious peroration. 'They had their meteorological chart. I *do* blame them. No right to take chances. From beginning to end they had troubles before they put off, but on account of public opinion, they did not turn back.'

Only Dr Archibald Fleming, the trenchant leader of the Church of Scotland, of the many preachers that Sunday who spoke of the R101, came near to matching that criticism. 'Whatever some of us may think,' he said from St Columba's on Pont Street, London, 'of the megolomania which has been the obsession of mankind from the days of Ninevah and Tyre and the Pyramid builders of Egypt down to the present age, no one can doubt for a moment the splendid heroism and devotion to duty of those who had charge of the airship and who have lost their lives in it.'

Next day, just as Mrs Radcliffe was about to leave London to see her husband, news came that he had died. Church's father and fiancée actually reached the nursing home, to be met by a nun who took their hands and told them sadly that Samuel had gone.

The bodies and survivors were brought back to England across the stormy Channel by the destroyers *Tribune* and *Tempest*, after a salute of 101 guns and a solemn procession led by Spahis on horseback and the 51st Infantry had taken place in the square at Beauvais, while forty-eight French Air Force planes flew overhead to honour the forty-eight dead.

At that same time, another seance with Eileen Garrett was taking place. A journalist called Ian Coster was researching the possibility of getting in touch with the spirit of Sir Arthur Conan Doyle, who had died in July. Harry Price, the well-known debunker of mediums, had invited him to come to his National Laboratory of Psychical Research at Queensberry Place.

Eileen Garrett was not told the purpose of the visit. She sat in an armchair, pinched her nostrils, began to breathe heavily. Then she began to talk very quickly in a strangely accented male voice, 'It is Uvani. I give you greetings, friends.'

This was her spirit entity. Price's secretary sat ready to take down all communications in shorthand, but none came. Then a Dr Feischner spoke. There was still no sign of Conan Doyle. Coster was beginning to think the seance was a waste of time when Eileen Garrett's eyes filled with tears.

'I see,' said the voice of Uvani, 'I see someone called . . . Irvin . . . no, he says . . . not Irvine . . . Irwin . . . he spells it I . . . R . . . W . . . I . . . N.'

There followed a jumble of sentences spoken by a faint voice very quickly in great distress.

'Must do something about it . . . apologize for coming . . . the whole bulk of the dirigible was entirely and absolutely too much for her engine capacity.'

The voice became stronger.

'Engines too heavy . . . useful lift too small . . . this new idea of new elevators totally mad . . . elevator jammed.'

Coster was surprised to hear technical jargon that meant little to him fall naturally from the medium's lips. Yet it no longer seemed to be a woman who was talking. The voice was staccato, male and fervent. When it was described later to people who knew Irwin, they said that was how he spoke.

'This exorbitant scheme of carbon and hydrogen is entirely and absolutely wrong . . . flying too low altitude and could never rise. Disposable lift could not be utilized. Load too great for long flight . . . cruising speed bad and ship badly swinging. Severe tension on fabric that is chafing.'

The voice paused. The listeners held their breath.

'Engines wrong. Too heavy. Cannot rise. Never reached cruising altitude. Same in trials. Too short trials. Airscrews too small. Fuel injection bad and pump failed. Cooling system bad. Bore capacity bad . . . fabric all water-logged and ship's nose down. Impossible to rise, elevator jammed. Almost scraped the roofs of Achy . . . kept to railway. At inquiry to be held later, it will be found that the superstructure of the envelope contained no resilience, had far too much weight. I knew then that this was not a dream but a nightmare. The added middle section was entirely wrong. It made strong but took resilience away and entirely impossible. Too heavy and too much overweighted for the capacity of the engines.'

Meanwhile, the bodies lay in state in a flower-decked Westminster Hall while a hundred thousand people filed by. There followed a memorial service at St Paul's, attended by HRH the Prince of Wales, before the bodies were again taken in solemn procession through the packed London streets to Euston station.

Viewing the scene from Tavistock Square, Virginia Woolf wrote in her diary for Saturday 11 October 1930:

. . . the fifty coffins have just trundled by, in lorries spread rather skimply with Union Jacks – an unbecoming pall – and stuck about with red and yellow wreaths. The only impressive sight was the rhythmic bending backward slow march of the Guards – for the rest, the human face was often pock-marked and ignoble. The poor gunners (?) looked bored and twitched their noses; the crowd smells, the sun making it all too like birthday cakes and crackers; and the coffins concealed too much. One bone, one charred hand would have done what no ceremony can do, and the heap of ceremony on one's little coal of feeling presses uneasily. I refer to the burial this morning of the 48 'heroes' of the R101. But why 'heroes'? A shifty and unpleasant man, Lord Thomson, by all accounts, goes for a joy ride with other notables and has the misfortune to be burned at Beauvais. That being so – we had every reason to say Good God, how very painful – how very unlucky – but why all the shops in Oxford Street and Southampton Row should display black dresses only and run up black hats, why the Nation's every paper should be filled with nobility and lamentation and praise, why the Germans should muffle their wireless and the French order a day of mourning and the footballers stop for two minutes silence – beats me and Leonard and Miss Strachan [clerk to the Woolfs' Hogarth Press].

Such a lack of compassion at the sight of human pain and frailty from a writer world-renowned for her sensitivity shows just how difficult it is to 'see' and 'feel' what is actually in front of our eyes, let alone what may still be subliminally visible and audible. Virginia Woolf showed herself there as deaf and blind on that wavelength as the captain of the Convair to the ranges of the mountains below him, though his co-pilot had already picked up the tremors. In the air are millions of stimuli clamouring for our attention, but we erect a perceptual defence against them. Our breadth of understanding is too narrow to encompass all but a few, the fire in 'our little coal of feeling' dies, so we cut the others out of our perception by denying they exist.

From Euston, the funeral train proceeded to Bedford, where again the streets were lined with people. Headed by a procession of the RAF carrying reversed arms and the Prime Minister's son, the Bedfordshire and Hertfordshire regiment with their band, dignatories of the county and cars conveying the relatives followed the coffins to Cardington churchyard where a huge mass grave had been dug. There, under the shadow of the vast hangars to the south where R101 had been born, most of her crew and all her passengers were buried.

A public memorial fund, intended mainly to provide for the dead's dependants, after all the interest and sympathy and sorrow barely covered the cost of the massive marble slab that was put over the grave. Mrs Radcliffe, who had been on the point of visiting her husband in hospital at Beauvais when she was told he was dead, received enough from the fund 'to buy a pair of eyeglasses'.

Sixteen days after the funeral, the inquiry was opened by the president, Sir John Simon, in the great hall of the Institute of Civil Engineers.

So important was its outcome regarded that the government assigned the Attorney-General and the newly appointed Solicitor-General, Sir Stafford Cripps, to the case. The assessors were Colonel Moore-Brabazon, who had obtained the first British pilot's ticket in 1909, and Professor Inglis, of the Engineering Department at Cambridge.

In spite of what Scott had said about bearing the responsibility for the flight, the spotlight naturally turned on the man designated as captain. The standard was that a captain such as Flight Lieutenant Irwin and the Minister in charge of a programme like Lord Thomson must accept the blame for actions for which they are nominally responsible.

The influential *Flight* magazine had stated that 'an officer of the Royal Navy who loses his ship is, if he survives, court-martialled and we feel it right that it should be so. It may be that in the case of aircraft accidents, we have been too squeamish'.

Major Teed, a friend of Barnes Wallis and trained both in law and engineering, volunteered to appear for Mrs Irwin to protect the dead Irwin's name. Significantly, no other officer or passenger was legally represented.

In the body of the court, Major Villiers, who had met his friend Brancker at Henlow, was in charge of a silver model of the R101 hanging from the ceiling. Three days before, sitting alone by the fire at his home, he had had a strange feeling that someone was there beside him. In his mind, it was as though Irwin's voice was saying, 'For God's sake, let me talk to you!'

Not previously interested in the paranormal, he consulted a friend who advised him to see a medium. There was a lady with an excellent reputation that his friend not only recommended but arranged for Villiers to see at 7 p.m. that Friday.

The Attorney-General opened the proceedings by saying that it was essential that every possible fact that could throw light on the disaster should be revealed. But as the extraordinary history of the R101's short life was unfolded, it was soon evident that this would not be the case.

'Who gave the authority to shorten the final test flight to only sixteen hours?' Sir John asked. No one knew. 'But surely there was a report of that trip?'

The Attorney-General said there was none in existence.

On his table Sir John kept a rough notebook in which he jotted down such private thoughts as '*there must be a report*' and '*who gave the order to go*'.

Teed told Simon that Irwin kept a carbon-copy manifold book in which he drafted all his reports. At least the carbon copies would be in his desk at his Cardington office. But the last test flight report could not be found. Air Ministry letters, reports and files had mysteriously disappeared.

Letters came into the court from the public. A bomb had been seen going on board – clearly the nine-gallon keg from the Bridge Hotel. The oldest magistrate in Bedford had reports of the crew's drinking habits. In this connection, an engineer of the reserve crew who was at the departure reported that two of the officers were drunk when they went on board. A colonel from Bedford wrote in to say 'one at least of the officers' – clearly he meant Irwin – 'did not expect to return'.

Sir John expostulated that he found it 'very difficult to believe there is no record of that final test flight'. Even the Attorney-General admitted it seemed 'very remarkable'.

Irwin's conscientious report of the Hendon flight *was* found. 'What strikes me very much,' Sir John went on, 'is that Captain Irwin evidently gave a most detailed report of the ship during her flights in June, and I am still very anxious to know whether there is not more material somewhere which shows how the ship behaved on the final test flight.'

The days went by, and Sir John was still complaining about missing documents. 'I very much object to the piecemeal production of relevant documents. Here and now I call for the production *in one group* of all Air Ministry and Royal Airship Works minutes which might conceivably refer to the cause of the R101 disaster.'

The court examined various causes.

Did R101 break up? Fragments of frame were discovered far from the crash. Was it the elevators? An elevator cable was found fractured, but tests showed the break occurred after the fire. Yet if that was so, why were some pieces of fabric only inches away quite unburned?

Was the airship turning round to go home? The direction of the crash was 225°, but the ship had been flying on 175°. Was the altimeter misreading? Was it the weather?

All were dismissed. Sir John was now concerned about a report Irwin had written on the gas-bags – also missing.

That Friday, Villiers had a seance with Eileen Garrett.

He had not told her his name or his interest, but once again Irwin's Irish staccato voice, remembered from his duty visits to Cardington from Air Ministry, came through.

'. . . one of the struts collapsed and caused a tear in the cover . . . the rush of wind caused the first dive and then we straightened again and another gust surging through the hull finished us . . . the forces she can be subjected to in bad weather and wind currents are too strong for the present system of calculation.'

Villiers was so impressed that he had a further seance on Sunday. Now he was told of a crew conference before departure, of the desire to postpone the flight, of missing documents.

He took long-hand notes, filling them into coherent form next day. He had an excellent memory, but there were undoubted mistakes and errors of fact. In all, he had seven sessions with Eileen Garrett.

Certainly the inquiry, which was meanwhile continuing, appeared to be getting even less close to what could possibly have happened than the spirit voices of Irwin, Scott and Brancker through the medium.

Sir John was still complaining about missing documents. 'Flight Lieutenant Irwin indicated in July that he had written a report on the gas-bags.'

A voice interrupted to say that a report had been found at Cardington.

'For the best part of a week,' Sir John thundered, 'the court has been asking for this report! It is a very important document. I have emphasised again and again that the court wants *all* information it can get . . . and immediately.'

Harry Price answered his call by sending him a transcript of the seance held weeks before with Eileen Garrett at his National Laboratory of Psychical Research. He had closely examined all the statements made by the spirit Irwin, and many of them appeared now to be coming out in the inquiry.

Sir John did not follow up Price's letter, but he did see Villiers and sent one of his secretaries to search Cardington for the missing documents reported in the seances. None were found.

Why the airship ever went off at all was troubling him. Was it because of Thomson's influence?

The Attorney-General declared, 'Not one of those very experienced officers – Irwin, Scott and Colmore – suggested to Lord Thomson that it was desirable that the flight to India should be delayed and that further tests should be made.'

Teed wrote to Sir John, informing him that as far as Irwin was concerned, the Attorney-General was wrong.

Irwin's gas-bag report, revealing the hundreds of holes caused by chafing, was at last produced.

Sir John's reaction was: 'These documents have come into my hands in the last half hour which for reasons I do not understand Cardington have not been able previously to produce. One is a letter from Richmond to Scott saying that he has calculated the loss of lift to be one ton per square inch of opening in twelve hours – a somewhat startling result. The second is a letter from the Cardington Maintenance Inspector, Mr F. McWade, addressed to the Secretary of the Air Ministry stating that the gas-bag situation was very serious, since the points of fouling throughout the ship "amounts to thousands". He goes on to say that until a proper remedy is undertaken he could not recommend the extension of the present permit to fly or the issue of any further permit or certificate, Why were these documents not produced before?'

Nobody answered.

Sir John asked McWade, 'What would you have done if it had been left to you whether the ship should have a certificate of airworthiness?'

'Had it been left to me, sir,' McWade replied, 'I'm afraid they would never have got it.'

Booth, captain of R100, gave evidence that the decision to leave was biassed by the imminence of the imperial conference. 'Irwin told me after the insertion of the bag that he hoped they could have a trial flight of thirty-six to forty-eight hours at a reasonable cruising speed under bad weather conditions so as to test the ship thoroughly.'

'Have you any reason,' Sir John asked him, 'to think that after her calm weather flight of sixteen hours, Irwin changed his previous view that a more elaborate trial in bad weather was needed?'

'No,' Booth replied, 'I have no reason to think that he changed his mind.'

Throughout the inquiry, Irwin was shown to have been repeatedly right. But the strain was telling on Mrs Irwin and, in a state of half collapse, she told her counsel that she would have to go away.

The court had now heard all the evidence and had broken up for a three-week recess to consider it.

Only now – and not in open court but in a letter from Teed to Simon – did the fact of Lord Thomson's extraordinary phone call to Irwin come out:

. . . Irwin had what may be described as very sound views on the trials. However, it is not before the court that he was telephoned by Lord Thomson who 'strafed' him for being an 'obstructionist'. Mrs Irwin, whose nerves are in a very, very bad state, has under medical advice started for the Cape. However, I am informed that Irwin did in fact mention the episode of the Secretary of State's conversation with him to his brother-in-law. Would you like this followed up . . .?

Sir John replied two days later to say he did not want Teed to address the court, adding, 'I may say that I am personally completely satisfied that there is no possible grounds to cast any reflection on him.'

Teed then telephoned Simon, suggesting he got a short written statement from Irwin's brother-in-law.

Sir John replied, 'Of course Lord Thomson may "strafe" a man in a very cheery and yet emphatic manner and it would be a very grave injustice to Lord Thomson's memory to make too much of that.'

Teed agreed that it would be difficult for a junior officer to turn up an attitude of refusing to fly with the Secretary of State and it would not 'in all probability have stopped the vessel from going when it did, for there were at least two and possibly three alternative captains, one of whom would have been prepared to take a chance which might well be associated with an award.'

The long overdue knighthood for Scott was implied.

Matters were left as they were. Lord Thomson's telephone call was hushed up. When the inquiry met again, little was added, though Simon publicly denied two rumours: one to the effect that there had been a woman aboard R101, and the other that crew members were drunk at the time of departure. He ended with a panegyric – perhaps guilt-inspired – on Irwin.

Four months later, the report was produced. The cause of the disaster it stated, could have been many things, but the most likely was that a gas-bag forward had suddenly burst, bringing the nose too far down for the coxswain to control it.

There was no criticism of Lord Thomson or the Air Ministry or the Royal Airship Works.

Certainly no criticism was levelled at Irwin either, but neither were his lonely efforts to protect the ship praised, and the traditional implicit responsibility of the captain remained. To absolve him from this, the report would have had to state categorically that the government in the person of its Secretary of State and with the acquiescence of its Air Ministry had assumed full responsibility for the safety of the airship and all who flew in her.

If Irwin was to get his proper due, Lord Thomson's name would have to be blackened. But instead a road at Shortstown, where many of those who died had lived, was named after him.

Sir John complimented himself and his assessors with the words, 'We have reached the truth without offending anyone.'

He was pleased with its reception by the flying Establishment. Lord Trenchard, leader of the RAF and Lord Amulree, the new Secretary of State for Air, were delighted. Sir John wrote to Teed, 'I am pleased to receive a grateful letter from Miss Thomson, Lord Thomson's sister, saying how greatly our report had relieved the minds of herself and her mother. It is a trying business for these people to be landed into poverty when with a turn of the wheel, they might have been out with the new Viceroy of India.'

Simon invited Teed's views on the report. 'To ask for my opinion,' the Major replied, 'is analogous to Sir Edwin Lutyens asking a plumber's mate what he thought of New Delhi . . . it is excellent from a technical point of view while from the human aspect, it is free of the slightest trace of harsh criticism.'

'One of the things,' Sir John wrote on 13 April 1931, 'about which I was particularly pleased is that Irwin's record and reputation stood out beyond challenge. Poor Mrs Irwin must get such consolation as this affords.'

She got precious little else – £200 and a £100 pension. The other government awards to the widows and children were equally ungenerous. Mrs Colmore received a gratuity of £450, a pension of £180 and £48 a year each for her children. Mrs Richmond, Mrs Scott and Lady Brancker each received an annual allowance of £300. Church's parents were given £100. The engineer Mason's widow was given £600, £62.10.9 pension and £13.14s. annually for each of her four children. All the other awards were in the same proportion.

The court broke up. Sir John warmly thanked Professor Inglis and wrote to Moore-Brabazon, 'My dear Brab, it has been great fun doing this with you. If only I could play golf as well as you do, the world would be a pleasant place. I spent a week at North Berwick and hit the ball a long way off the tee.'

North Berwick – where an Irish airship captain called Herbert Carmichael Irwin met the pretty local girl called Olive Teacher he was to marry.

The mystery of the R101 remained.

Years later, Moore-Brabazon admitted that the assessors had never really got to the true cause of the R101 disaster. The inquiry was now regarded as a gigantic Whitehall whitewash to save the memory of Lord Thomson and protect the Establishment.

Gradually it became apparent that the seances through Eileen Garrett had got closer to the truth than the expensive inquiry. R101 *was* too heavy, her lift *was* too small, the elevator *might* have jammed, a mixture of carbon and hydrogen from the gas-bag for use as fuel *was* being tested at the time and proved useless, she *did* fly too low and had difficulty rising, her trials *were* too short, she *was* difficult to trim, the added middle section was entirely wrong, taking away 'resilience'. A supply officer at Cardington, William Charlton, who had worked on R101, corroborated these findings.

Then an engineer called Spanner published two volumes on *The Tragedy of the R101* which also supported some of the causes outlined in the seances. He considered that the airship had 'hogged' – that is, its tail was in different weather to its nose, one part caught in a gust, the other in a lull, and it became deformed. Price claimed that Spanner supported the seances in stating that R101 broke her back through structural weakness, suddenly encountering a head-on gust of about 66 miles an hour, which caused a rapidly increasing downward motion. To correct the dip, the elevating fins were moved to a fully up position (they were found so in the crash) and the airship began to rise. While this was taking place, the lower girders gave way in the new bay inserted in the middle of the airship.

After the tragedy, R100 was flattened with a steam-roller. All work on the mammoth R102 was stopped. Britain turned her attention from airships to aeroplanes.

Major Villiers continued as an unestablished officer at the Air Ministry before retiring with no pension. His eyes began to trouble him and for thirty years he was blind. The Garrett seances totally convinced him of life after death, and it was the spreading of this conviction which guided and up-held his life. He died in 1981 at the age of 94.

Sir John Simon went on to high office, becoming Foreign Secretary, Home Secretary and Chancellor of the Exchequer under prime ministers of a variety of political hue. In 1938 he became one of the appeasers of Munich. Later he was raised to the House of Lords as Viscount Simon.

Moore-Brabazon became Lord Brabazon of Tara and had the first jumbo named after him – a huge Bristol aircraft built at vast expense and never put into production.

Dowding, who flew on that last R101 test flight, became Chief of Fighter Command and undoubtedly saved Britain with his policy and direction of the Battle of Britain. He later became interested in spiritual studies, and wrote two books on the psychic world.

Ramsay MacDonald was devastated by the loss of Thomson. 'When I bade him goodbye on Friday,' he wrote in his diary, 'and looked upon him descending the stairs of Number 10, that was to be the last glimpse of my friend, gallant, gay and loyal. No one was like him and there will be none. Why did I allow him to go? He was so dead certain that there would be no mishap.'

The Prime Minister had been introduced to Princess Bibesco by Thomson, who brought her to Chequers to dine at MacDonald's invitation. After the disaster, they met regularly. From then until his death, he wrote to her almost every week – wistful, sometimes faintly flirtatious letters. He teased her about her cat, her love for foreign travel and her Chanel scents, insisted gravely on the superiority of his Presbyterianism over her Catholicism and then signed his letters with a drawing of three swans, as a tribute to the swans that swam on a lake beside her home in Romania.[1]

He also became interested in spiritualism when a medium wrote to him. He replied that he knew little about such things, though he was open-minded and prepared to believe. That was the start of an interchange that lasted several years.

Eileen Garrett, the medium on whose integrity much of the mysterious story of the R101 hinges, went to America where she established herself as a medium of great power and integrity. She continued to search for the meaning of her gift, and freely submitted to investigation by Duke University and elsewhere.

She also formed a close friendship with Mrs Frances Bolton, the congresswoman from Ohio. Mrs Bolton, a Republican, was unsure of how to vote in the 1941 Lend-Lease Act. Most of her fellow Republicans were voting against Roosevelt's measure to help hard-pressed Britain with much-needed supplies. In a session with Eileen Garrett, Uvani's voice came through, and under his guidance, Mrs Bolton gave the casting vote for President Roosevelt's Bill, a most significant and tide-turning measure.

Eileen Garrett remained until her death one of the few mediums never exposed. She started a successful publishing house. And with the help of Mrs Bolton, she started the Parapsychology Foundation in New York which has attracted interest and encouraged experiments all over the world.

The arguments over the disaster still go on – for the R101 is a mystery within a mystery – a Greek tragedy. Embedded in the enigma of the main event are smaller mysteries and coincidences: the predicted fate of Sir Sefton Brancker and its rehearsal at Beauvais two and a half years before; the escape of Binks and Bell; Mrs Garrett's clairvoyance; the disappearance of relevant documents; premonitions of disaster; the conspiracy of silence between Teed and Simon over Lord Thomson's telephone call to Irwin.

The hangars at Cardington still remain – gaunt pyramids jutting above the flat Bedfordshire skyline, now in none too good repair. Pigeons scratch and roost in that vast roof, dropping pieces of metal like bombs on the floor beneath, so that thick plastic helmets have to be issued to visitors. Balloons are nevertheless still built there. And on 25 September 1981, a new airship, the helium-filled Skyship 500, was launched by a private company and made its first flight.

Officials say that the sheds are so huge that by some freak phenomena they have their own weather. Sometimes they are foggy when the weather is clear outside, and fine when outside it is wet and windy. Once it even snowed in Number One shed, where the R101 was built, when there was no snow outside.

Of those men and women of the R101 age, few remain. One of the last is Sir Victor Goddard who served with distinction in the Pacific under Admiral

1 David Marquand: *Ramsay MacDonald*, Jonathan Cape, 1977.

Nimitz in the Second World War. Now he too, like Lord Dowding was, is deeply involved in spiritual and religious studies. A few years ago he tried to contact the spirits of his dead airship friends, or at least perhaps discover that they had never revisited this earth at all.

He invited a distinguished medium to accompany him to Cardington. Together they walked into the cold eerie Number One shed over the concrete floor. Above them towered the iron-raftered roof. Footsteps and voices echoed in the vast emptiness. It was the perfect atmosphere for a spiritual visitation. But the medium experienced nothing. No voices, no visions, no sensations, nothing.

Disappointed, they returned to the hotel. It was while the medium was alone in her room that the visitation came. Effortlessly and without preamble, as if he moved freely now and without struggles, the spirit of Irwin materialized.

The medium had no doubt it was he, Irwin. Irwin, he said, I-R-W-I-N, as he had done at the seance all those years before.

Gently he asked that no further attempt be made to contact himself or any of those who had perished in the R101 disaster. He gave her to understand that Lord Thomson had realized he was wrong, that he with all the others was now doing useful work.

But now . . . Irwin conveyed to the medium a profound feeling of love . . . they needed to be left in peace.

And the physical remains of Bedfrod's R101? The huge twisted girders, that web of metal? Shipped back on the Cooperative Wholesale Society's freighter, *SS Friendship*, they were then sold as scrap to the highest bidder.

That bidder was the Zeppelin Reederei at Frankfurt Maine where, when cleaned of the residue of the holocaust, they were incorporated into the frame of Frankfurt's pride – the Zeppelin *Hindenburg*.

The *Hindenburg* became Frankfurt's ship as certainly as the R101 belonged to Bedford. Herr Krebs, Frankfurt's Chief Burgomaster, had presented Captain Eckener, who had stepped into Count Zeppelin's shoes as the world authority on airships, with an inscribed silver cup after the *Hindenburg*'s first flight to Lakehurst.

Twice as big as the *Graf Zeppelin*, sixteen cells filled with hydrogen were contained in his (the Germans referred to their airships in the masculine) 803.8 feet envelope, itself made of a gas-tight skin like photographic film cemented between two layers of fabric. Modern in every detail, he was Germany's pride of the skies as he began what looked like being a most successful and triumphant career. On Monday 3 May 1937 he had already completed ten round trips across the Atlantic and another to Rio, carrying in comfort and safety over a thousand passengers, when he slipped the mast that evening.

Like the R101, he was threatened not only by wind and weather, but also by the currents of a worsening political situation. Adolf Hitler, who had ousted the ship's patron Hindenburg, was now Führer and Chancellor. The persecution of the Jews was under way. Nazi storm-troopers swarmed the streets. There were gigantic hysterical rallies at Nuremberg. German and Italian troops on the one side, Russian on the other, fought their rehearsal for war in devastated Spain. It was known to the Zeppelin Reederei at Frankfurt (equivalent to the Royal Airship Works at Cardington) that Hitler did not approve of the airship's name. Indeed he never used it. He was displeased that Captain Eckener had not destroyed the bust of Hindenburg in the foyer and replaced it, as any loyal captain would, with one of his beloved Führer. Hermann Goering, an aeroplane pilot himself, was also less than enthusiastic about airships, though he saw the propaganda value of this fine example of German technical skill.

It was also suspected that though most of her crew were Nazi party members, their interest was nominal. The day before *Hindenburg* sailed, there was a Hitler Youth Rally at the Olympia Stadium. Dr Goebbels, the Propaganda Minister, had requested the airship's presence in the sky overhead. On the excuse of the weather, the airship was unable to comply.

There was friction between the Air Ministry and the Reederei. It was suspected that some of the crew might be seduced by Communist or anti-Nazi agents. One indeed was known to be a Communist party member. And as if to support this fear, rumours were rife that the airship would be sabotaged. The world was moving inexorably into the era of the bomber, the terrorist and the hijacker. Bombs had been found on the zeppelins *Nordstern* and *Bodensee*. One had luckily been spotted on the *Graf Zeppelin*. The rumours had their effect. Max Schmeling, the boxer, had been booked on a *Hindenburg* flight, but went instead by sea. A journalist similarly booked received a telephone call from an anonymous man warning him not to go.

Though the Air Ministry derided them, the threats of sabotage had an adverse effect on the 1937 bookings, though on the return of this first trip, they were fully booked with passengers coming to Europe for George VI's coronation. Then in April, the German Ambassador in Washington, Dr Luther, received a letter from a woman in Milwaukee, Kathie Rauch, which somehow impressed him with its urgency and sincerity. Why it should have done, he did not afterwards know. Certainly it was only one of many he had received She wrote in uneducated German. Apparently, her lodger had had a vivid dream of the *Hindenburg* burning.

Begging the Ambassador to forward her letter to the Zeppelin Company, she went on, 'They should open and search all mail before it is put on prior to every flight . . . the Zeppelin is going to be destroyed by a time bomb during its flight to another country . . .'

So convinced was Dr Luther that he forwarded the letter immediately to the Reederei who in turn sent it to Dr Lehmann, once the captain of the *Graf Zeppelin*. The result was that Captain Lehmann decided to accompany this first flight of the season. He was one of the most experienced airline captains in the world. Furthermore, he sought some worthwhile task to help assuage his grief after the recent sudden death of his 21-year-old son.

The second result was that everyone and everything going aboard *Hindenburg* that wet evening was subject to minute scrutiny. Because of the low bookings, several of the passengers had complimentary tickets, whilst among the crew there were sure to be one or two whose primary job was to observe the behaviour and sound the politics of the rest.

As with R101, all passengers and crew were searched for any inflammable material, especially matches or dry cell batteries before they left the hotel where they assembled. In the customs shed the men who meticulously opened and searched the cases were in reality Nazi police. Here everything was weighed. Yet with all the cargo being loaded, most observers agreed that it would have been comparatively easy for someone to slip on board him as he rode at the mast with his hatchways open. While the passengers embarked, the rain cleared and a pale watery sun came out.

Just before the gangway was stowed, a late passenger who had not been through the search at the hotel arrived in haste by taxi. He was carrying a parcel and a case. No one saw them being searched.

Shortly afterwards, Captain Pruss gave the order to slip the mast. To the strains of *Deutschland uber Alles* and the *Horst Wessel Song*, *Hindenburg*, or LZ129 as Hitler preferred to call him, set course for Lakehurst, New Jersey.

The night was at first clear and starlit, but there were build-ups of cumulo-nimbus over the English Channel, and further out over the Atlantic the perennial head winds that hamper the westbound crossing. *Hindenburg* coped smoothly with wind and cloud, as if to contradict Dr Eckener's assertion that he had been born under an unlucky star.

By 10 am on 5 May he was 300 miles off St John's Island, New Brunswick. The mood among the passengers, which at first had been subdued, now lightened. Captain Lehmann, however, continued sad and withdrawn. As well as his grief over his son, he was worried about his wife. Stricken with grief, she was also concerned by rumours of sabotage. Frau Lehmann had consulted a clairvoyant. This woman had told her that her husband would perish in an airship fire, thereby adding to her terrible distress. The clairvoyance Lehmann could unhesitatingly dismiss, but he could less surely dismiss the letter from Frau Rauch which he still carried in his pocket.

Apart from his personal worries, there was the technical worry that the *Hindenburg's* gas-bags were filled with hydrogen, a gas much more inflammable than the helium which filled American airships. Distrust of the Nazis and fear of war prevented the USA from selling helium, of which they were the chief producers, to Nazi Germany. The Führer's behaviour was not likely to convince the Americans to reverse their policy.

Captain Lehmann enjoyed the confidence of the German Air Ministry. He knew that at least one of the passengers was suspect, that one of the riggers was friendly with a former Communist, that however minutely the aircraft was searched, however meticulously the rules about no unaccompanied passengers along the catwalks, no matches, no batteries, no flashlights, there was always someone prepared to smuggle something. And in the folds and loops and crannies of this great ship, always somewhere that something might be hidden. Always, someone who would like to see that hated Nazi swastika on her tail-fin utterly destroyed.

Still out over the Atlantic, the headwinds increased. Electrical storms interfered with the radio. Yet comparatively smoothly, though much more slowly, *Hindenburg* nosed forward.

Then down below they glimpsed icebergs and the coast of Newfoundland. On Thursday morning they were over Boston lightship. All the stories of doom were manifestly disproved. And though weather conditions at Lakehurst were poor, the passengers decided the trip had been too short.

It was in fact lengthened. A cold front and polar air were warring violently with warm moist air from the south. There were heavy build-ups of anvil-topped thunderclouds, lightning and downpours of heavy rain.

As *Hindenburg* cruised slowly towards Long Island Sound, Hugo Eckener was visiting the sculptor Ambrosi at Graz. There the artist insisted on showing him a sculpture of Icarus plunging into the sea.

The weather over New York began to deteriorate. Arriving over Lakehurst, the captain sent a signal 'Riding out the storm'.

They rode it out through intermittent storm cloud and fragile sunshine until *Hindenburg* received the message that the thunderstorm was moving away. Despite this, it was almost half past six before Captain Rosendahl, in charge of the landing, radioed that conditions were favourable for landing. The mooring cone had been in place since morning and the soaked ground crew were once more ordered out.

Everything aboard *Hindenburg* was still shipshape, except for one minor point. While making a routine inspection, the chief rigger noticed something different about number 4 gas-bag, something he could not quite discern, something that damaged it, something he should further investigate. But at that point, Captain Pruss, nudged by Rosendahl, committed himself to landing, and a high landing at that.

Everything was now bustle, and all the crew were ordered to their landing stations. In over Lakehurst came *Hindenburg*. Rosendahl had been right. The thunderstorm had moved off. The cloud ceiling lifted. Captain Pruss weighed off, valving gas to get him in balance for the locking-on. Descending slightly tail-heavy, water ballast was released. Crewmen were ordered up front till he was straight and level.

At that point there was a sudden shift of wind necessitating a turn (just before the R101 crashed, there had been a similar windshift). One observer remarked, 'Lehmann must be in a hurry to get down.' Then the manila nose-handling lines were dropped and grasped by the handlers.

At that point, held by the port bow rope, two observers noticed a fluttering of the outer cover on top of the port side, while a watching engineer saw a ball of flame underneath the ship. To an engineer on the crew who saw the same phenomenon at close quarters it looked like the switching on of a flashlight. To him it came from the catwalk near number 4 gas-bag. It was accompanied by a low sound like the lighting of a gas oven.

Another observer from the ground saw a faint puff of blue smoke on top of the envelope. 'A small burst of flame,' Captain Rosendahl described it. Only a few hundred feet up, passengers were laughing and waving from the windows when, as *The Times* reported, 'a bomb-like explosion sent out clouds of red and billowing smoke'.

Next moment the crowds were shrieking, 'Run for your lives!' as a column of yellow flame shot up and the airship erupted.

'Get out of the way!' commentator Herbert Morrison shrieked into his microphone. 'My, this is terrible! Get out of the way, please! It's burning, bursting into flames and falling on the mooring mast and all the folks . . .'

The muffled boom of the airship going up was heard fifteen miles away. Yet ironically the people on the ground were aware of what was happening before those on board. Survivors described a sudden stillness, others a sound no louder than a rifle shot, followed by an explosion and 'a mass of shrieking crying people' as flaming wreckage and bodies descended everywhere.

There was a terrific explosion which was seen or felt by all the spectators. Then flames shot upwards and the *Hindenburg* burned 'like tinder'. The buildings reflected light as the fire spread around the field and, after swaying for a moment, the long grey cigar shape collapsed with a terrific impact into a holocaust of fire.

Harry Wellbrooke was one of the ground crew who ran for their lives to get out of the way of the blazing wreckage. He said, 'We got three bodies from the stern of the ship. All were burned beyond recognition, but one, whose features were unrecognizable, was still breathing. The clothing on all three was burned to a cinder.'

At Graz, Eckener was woken at 2 am. Before he was told, the first thing that came into his mind was that sculpture of Icarus descending.

Thirty-two lives were lost, and several others, including Captain Pruss and Captain Lehmann, were severely burned. An inquiry was at once convened by the Secretary of Commerce, while behind closed doors and under conditions of strictest secrecy, the German officer survivors began their own inquiry.

Pre-empting both these inquiries, Adolf Hitler gave credence to a power he habitually refuted by declaring to the world that the disaster was 'an Act of God'.

In this he was supported by Goering and Dr Luther, the German Ambassador. While Frankfurt mourned, a directive was received from the German Air Ministry that the officers of *Hindenburg* were not to enquire too deeply into the cause of the disaster, nor were they to volunteer information to the inquiry.

A conspiracy of silence once more hampered an airship inquiry.

Answers that seemed deliberately vague were given. The job of the inquiry was further hampered by the depradation of souvenir hunters who stripped the wreckage like locusts, no doubt taking with them valuable clues.

Before he died of his burns, Captain Lehmann said he believed it was sabotage. But Eckener stated that the ship had made a 'very large turn to port' in order to head into wind which had shifted to the southwest. 'It is possible,' he went on, 'that with such a sharp turn – which is preferably to be avoided – a bracing wire broke . . . and the recoiling end of the broken wire could have torn a hole in the cell.'

The R38 broke up over the Wash because of a too sharp turn first one way, then the other. The captain of the *Hindenburg*, knowing of the warnings of disaster, must have been under immense pressure to put down. On top of that, there is a psychological condition known as the Zeigarnik Effect which is the tension caused by uncompleted tasks – a stress situation which affects most people.

At the time of the disaster, preparations were being made by Imperial Airways and Pan American jointly to cross the Atlantic in flying boats. This they did, eliminating further need for airships on long-range routes – but that same tension to get down quickly was to remain a major cause of aircraft accidents.

If the cell was punctured, the hydrogen would rise and, together with air, form a combustible mixture in the vicinity of cells four and five, where it might be ignited by a brush discharge.

Eckener pointed out that the first appearance of a naked flame was on the top of the ship just forward of the upper vertical fin. Both the German and American inquiry reports supported Eckener, though the German one states, 'the possibility of deliberate destruction must be admitted'.

How the spark was produced that ignited the mixture was a subject of considerable controversy. There was certainly lightning that day – but none had struck the ship. St Elmo's fire was put forward as the culprit – the brush discharge that makes the windscreen of aircraft alive with the phosphorescent snakes of electric rain and which I used to watch for hours at a time over the Atlantic, harmless rings of blue or yellow and silvery light ringing the propellers.

'Wet ropes' were another culprit, but the ropes were proved to have been dry. A professor of engineering declared that there was nothing to support the theory of any electrostatic discharge whatever being responsible. Other experiments at last made an inflammable spark from laboratory-produced St Elmo's fire, though at a peak intensity.

Another theory was that the blue smoke seen on top of the envelope was caused by an incendiary bullet fired from the ground right through the aircraft (there had been incidents on both sides of the Atlantic of cranks firing at airships and aeroplanes). There was a rumour of a pistol being found near the scene with one bullet discharged.

What was not a rumour was either deliberately or accidentally downplayed. A policeman had found a charred bag among the wreckage, and sent the contents for analysis. These were found to be the remains of a small dry battery, of the type forbidden to be carried on board: a little wad of manganese, graphite and zinc oxide that could certainly produce a potentially lethal spark.

Like Captain Lehmann, Captain Pruss continued to believe it was sabotage. A device might have been put on board, timed to destroy the ship after the passengers had disembarked, and had misfired because of the airship's delay.

There was a theory that Goering himself, mistrustful of Eckener and disliking the whole airship project, had ordered the sabotage as he now ordered the silence at the inquiry. Certainly the Nazi hierarchy was capable of any madness and duplicity, as witness the Reichstag fire.

But how exactly was the disaster caused?

Was it a bomb? If there was no question of sabotage, why were some of the apartments of those on board searched by the Nazi police? Why was one of the riggers being watched?

Or in the light of new psychological knowledge, was it human error? Did Captain Pruss, tired after his long battle with the winds and already twelve hours late, worried about the constant threats, make too eager a turn to get down? Did a bracing wire snap and puncture a gas-bag? Did an electrically-produced spark ignite the escaping hydrogen?

The questions go on and on far beyond the inquiry, as they do with the R101. And in many ways, the *Hindenburg* was an echo of the R101.

Yet the *Hindenburg* had been specially designed to avoid the fate that overtook the British airship. There was hardly an inch of wood in the ship. The walls were all of balloon fabric. Everything was built of featherweight duralumin – but much of the metal originally came from the R101.

3

Through
the Looking-glass

$$\diamond\diamond\diamond\diamond$$

'It's all imagination' – that is the reason often given to explain parapsychology, telepathy, extra-sensory perception, ghosts, precognition, unidentified flying objects, astrology, coincidence, luck, fate, reincarnation, life after death, prophesy and the mysteries of Time and Space.

It was also the reason given to explain the impossibility of flying, of the steam engine, of radio, of television, of going to the moon and many other things that these days we take completely for granted. Anything new and mysterious the human being initially scorns as 'imagination'.

Yet imagination itself is mysterious. No one knows what it is. Do we somehow manufacture the pictures from inside ourselves? Are they new leaps forward into the unknown or are they compensations for deficiencies? Why do so many of them portray the future?

In 1865 Jules Verne in his novel *From Earth to Moon* imagined a flight of three astronauts from Florida in a spaceship called the *Columbird*, taking ninety-seven hours and thirteen minutes. The command module of Apollo XI was called *Columbia*, and this American spaceship left from Florida with three astronauts on board, taking ninety-seven hours and thirty-nine minutes. Verne's spaceship got into trouble and had to use its rockets to return to lunar orbit as did Apollo XIII – and both factual and fictional spaceships landed in the Pacific.

Imagination is often a derogatory term. Yet to be called 'a man of imagination' is a compliment.

One such man, Captain Balfour, the parliamentary Under-Secretary at the Air Ministry, went over to America in August 1940 to initiate the training of RAF pilots. While he was there he heard of the possibility of obtaining from Pan American three of the new giant Boeing 314 flying boats, the first commercial Atlantic aircraft. He immediately bought them for $1,050,000 each. He had no authority whatever from the British government to do so, and as soon as he returned to England, he was in hot water. Beaverbrook, the Minister of Aircraft Production, would not speak to him. The Battle of Britain had just taken place. Churchill wrote to him icily to say that his failure to consult 'has caused me a great deal of work and worry at a time when much else is happening'.

His worry and that of Balfour's would have increased considerably had they known what had been happening behind the scenes with the Boeing 314.

At 83,000 lb, the biggest flying boat operating, and carrying seventy-four passengers, it looked magnificent. It had double decks, a restaurant, a lounge, passenger bunks and a honeymoon suite in the tail.

It had sponsons or sea wings 'for lateral stability in the water' – yet initially was always on the point of capsizing and was extremely difficult to land. Each time it came down on to the water, it bounced, skipped and porpoised. One of them had crashed badly at Horta in the Azores. Even when taxiing, pilots were regularly dunking their wings in the water. Frantic efforts were made to correct the plane's faults. Boeing constructed two models and checked them in the testing-rig. Take-offs and landings having been found almost impossible, every sort of alteration was made to the hull, including the removal of sponsons altogether.

Then a man with imagination had the idea of fixing a block of wood to the model that extended the hull twenty inches further aft. There was no scientific or aerodynamic reason why it should make any difference at all. It had been a sudden brainwave, a burst of intuitive imagination.

The row over the Balfour Boeings, as they were now called, was still going on in January 1942 when the principal protagonist, Churchill, also a man of imagination, boarded one of Britain's three, the *Berwick*, to fly from Baltimore to Bermuda. He had been seeing Roosevelt and his farewell words to Harry Hopkins, the President's friend, had been, 'Now for England, home and a beautiful row!'

Captain Kelly-Rogers flew him down. An Irishman with service in the merchant navy and Imperial Airways, he immediately got on well with Churchill. The Prime Minister was also impressed with *Berwick*. He took the controls for a while and flew this ponderous machine of thirty or more tons in the air. Coming in over Bermuda harbour, he saw the battleship *Duke of York* waiting to take him to England 3500 miles away, and the thought struck him that it would be altogether quicker if he went home by flying boat.

On the ground, the high brass were against the idea. Marshal of the RAF Portal, Chief of the Air Staff, said the risk was 'wholly unjustifiable'. The Naval Chief of Staff agreed with his colleague. Max Beaverbrook, still Minister of Aircraft Production and the third protagonist in the flying boat row and still not on speaking terms with Balfour, was also strongly opposed.

Churchill insisted, clinching the matter by saying, 'Of course there will be room for all of us.' He would leave in *Berwick* at 2 pm next day, 15 January. The two Chiefs of Staff, Beaverbrook, Charles Wilson, the PM's physician, and Hollis, Secretary to the Chiefs of Staff Committee, would accompany him. As one of Kelly-Rogers' friends was to put it later, 'All the baskets in one egg.'

Next day Churchill admitted, 'I woke up unconscionably early with the conviction that I should certainly not go to sleep again. I must confess that I felt rather frightened. I thought of the ocean spaces, and that we should never be within a thousand miles of land until we approached the British Isles. I thought perhaps I had done a rash thing, that there were too many eggs in one basket. I had always regarded an Atlantic flight with awe. But the die was cast.'

After lunch, they all went out by launch in brilliant sunlight and boarded the flying boat. Fully laden with petrol, *Berwick* thundered over the water, lifting up over the low coral reefs and then heading east to Pembroke Dock in Wales.

It was smooth. The passengers passed an agreeable afternoon and that night had what Churchill called 'a merry dinner' in the luxury restaurant with good food and excellent wine. Hollis and the Chief of the Air Staff played picquet, and Portal also demonstrated card tricks.

By that time, darkness had fallen and they flew on in dense mist at around 7000 feet. From the engines, long yellow exhaust flames could be seen pouring back over the wing surfaces.

Beaverbrook sat up all night reading. Churchill retired to the bridal suite in the stern with large windows either side, and slept soundly for several hours.

He woke before dawn. The steward had thoughtfully warmed his shoes by popping them in the aircraft's oven. Dressing and going forward, he stood behind the pilots on the flight deck. He then became interested in the rubber de-icer boots inflating and deflating to break the slight rime on the leading edge.

Fog was reported at Pembroke Dock where a special train was waiting to take Churchill to London. The flying boat was cleared to Plymouth at 1500 feet, and altered course accordingly.

Just before ten o'clock, the co-pilot pointed through the windscreen and said, 'The coast.'

Looking down through misty conditions, Kelly-Rogers caught sight of the Mewstone, followed soon after by Stratton Heights and the RAF Station and Mount Batten where as usual he alighted perfectly on the water.

They had had an uneventful 3300-mile trip that had taken seventeen hours and fifty-five minutes.

Nevertheless, the responsibility had been heavy. When he said goodbye to Churchill, the captain remarked, 'I never felt so much relieved in my life as when I landed you safely in the harbour.'

Churchill was totally sold on flying boats. He told Balfour that he withdrew entirely his censure on him for having bought them. He then gave hell to Beaverbrook for having opposed the purchase, and Beaverbrook wrote to Balfour and apologized.

Churchill said the Balfour Boeings were 'fine ships'. They were indeed – now. A single wooden block as had been fitted on intuition alone to the model had also been attached to all the Boeing 314s. Now they were a pilot's delight, docile both on take-off and landing. *Berwick* and her two sisters operated continuously with BOAC during the war, maintaining regular services to Lagos and across the Atlantic, flying four million miles, carrying passengers, diplomatic bags, precious cargo and mails without accident. Churchill made several more trips across the Atlantic in them with Kelly-Rogers in command.

The Irishman had every reason to be pleased and proud. And then in 1950 the third volume of *The Second World War* was published, and Kelly-Rogers read with astonishment Churchill's account of his first air crossing of the Atlantic.

'After sitting for an hour or so in the co-pilot's seat,' Churchill had written,

I sensed a feeling of anxiety around me. We were supposed to be approaching England from the south-west and we ought already to have passed the Scilly

Islands, but they had not been seen through any gaps in the cloud floor. As we had flown for more that ten hours through the mist and had only one sight of a star in that time, we might well be slightly off our course. Wireless communication was of course limited by the normal war-time rules. It was evident from our discussions that we did not know where we were. Presently Portal, who had been studying the position, had a word with the Captain, and then said to me, 'We are going to turn north at once.' This was done and after another half hour in and out of cloud, we sighted Plymouth.

Churchill added that later he had learned that

if we had held our course for another five or six minutes before turning northwards we should have been over the German batteries at Brest. We had slanted too much to the southward during the night. Moreover the decisive correction which had been made brought us in, not from the south-west, but from east of south – that is from the enemy's direction rather than that from which we were expected. This had the result, as I was told some weeks later, that we were reported as a hostile bomber coming in from Brest, and six Hurricanes of Fighter Command were ordered out to shoot us down. However they failed in their mission.

One of the most heavily defended fortresses in Europe, with a balloon barrage, numerous anti-aircraft batteries and fighter squadrons, Brest was the main German U-boat base. To take the British Prime Minister over it by navigational error was a grave slur on the captain's reputation.

Captain Kelly-Rogers protested against this flight of imagination in a letter published by the *Daily Telegraph* on 15 April 1950. 'Mr Churchill,' he wrote, 'displayed exceptional skill in navigating the Empire through the uncharted seas of war, but in the course of this and subsequent flights over the Atlantic with me, I have to say he appeared to take no more than the layman's interest in the art of aerial navigation and was inclined to dismiss the whole thing as a bit of a mystery.'

One would have expected that would have been the end of it. But no – now came the 'corroboration' from the distinguished passengers. One by one, in their memoirs or biographies, they published accounts of the incident that remained in essence faithful to the Prime Minister's account.

Moran in his diaries said they had lost their way. Beaverbrook said they saw Brest through the clouds and declared it to have been 'a foolhardy journey'.

General Sir Leslie Hollis, in James Leasor's *War at the Top*, said they were within a matter of minutes of coming out into 'a clear sky over Brest, the most heavily defended French occupied port in the Channel. We were so low and so slow that the German gunners could not have failed to bring us down. I seriously considered tearing up my report and burning the whole thing in the aircraft kitchen rather than risk its capture.'

I first came on the story of Churchill's first Atlantic flight when I was writing a book, *The Water Jump*, about the air conquest of the North Atlantic. I wrote a couple of paragraphs, largely made up from Churchill's account since I thought that with such illustrious supporting witnesses there was no need for further corroboration.

However, before publication I did send the paragraphs to Captain Kelly-

Rogers in draft for his comments. He had been my line manager in Montreal when I was on the BOAC Atlantic route, and a very efficient and disciplined one he was. His initials were J.C. and, perhaps inevitably in his position, he was referred to as Jesus Christ. Various stories about him in this capacity circulated. Once he had landed a flying boat and tied up to a buoy in the middle of the harbour. As the motor-launch tender came alongside to pick up the crew, Jack Kelly-Rogers told them, 'You go on ahead in the tender. I'll walk.'

I remember shortly after I had arrived on the line there was some grumbling amongst my fellow first officers. What about I never did discover, but a round robin signed by a number of first officers, which I never saw, was sent to the management together with a letter listing a string of grievances.

Kelly-Rogers had all the first officers into his office. He read out the letter sentence by sentence, tearing each one into shreds before proceeding to the next. He then asked if any first officer wanted a posting to another line.

Nobody did. The Atlantic was the crack BOAC route and with station allowance and Atlantic pay, we received nearly twice as much as other BOAC pilots.

He then asked if anyone had anything else to say. Nobody did.

As we shuffled out through the door, he sent a Parthian arrow into our retreating backs. 'I'm running an airline, not a girls' school!'

This time he sent my Churchillian draft paragraphs back to me with one rude five-letter word beginning with B. He then told me the whole story.

Kelly-Rogers report, handed in after the flight, shows the route followed to Plymouth was totally normal and his time of arrival as predicted. Had it been as suggested by Churchill, he would have arrived at least half an hour later. Furthermore, the flying boat was under radar surveillance controlled by Prestwick during the later stages of the flight and, according to the control officer, was 'nowhere near Brest'. There was no record of the Hurricanes being scrambled.

It had all been sheer imagination – and imagination has its own hypnotic power under which the 'witnesses' fell. But Kelly-Rogers naturally felt the slur very deeply as he continued his distinguished aviation career. For Churchill's after-dinner story stuck, like all good fairy tales. The slipper in the original *Cinderella* was made of fur, not glass, but no one could possibly change it now. For some mysterious reason, we *want* one of our greatest prime ministers to have almost flown over the dragon's lair of Brest.

So it went on. In the obituary of Lord Portal, Marshal of the Royal Air Force in the *Daily Telegraph*, 24 April 1971, the story appears yet again:

Portal, who had been intently studying the map, abruptly instructed the pilot to turn north immediately. That was done, and half an hour later, England was sighted. Subsequently it was estimated that if the course had not been changed, the aircraft would have been flying over Brest within five minutes and at the mercy of the German batteries.

The borderline between the world we live in and the imaginative world is a thin one indeed. The other side of the Looking-glass is with us every day.

Quite what inspired Churchill's flight of imagination into the Looking-glass will never be known. Remembrance of his young days fighting in the Boer

War? Frustration that he now had to sit at a desk while others were getting all the excitement? Whatever it was, out he went, Walter-Mitty-like, on to the limb of unreality when all the time prosaically he was on the dull tramlines of a well-planned and perfectly normal flight.

In 1951, another man – this time a trained navigator – looked through the Looking-glass from inside out. He used his imagination the other way round. He imagined a series of events to keep him on the dull tramlines of a well-planned and perfectly normal flight when in fact it was anything but.

Close to midnight, he took off from Tripoli in a Hermes to fly across the Sahara to Kano in northern Nigeria.

The meteorological officer warned of the possibility of thunderstorms en route, and the crew were told that the Kano non-directional radio beacon was unreliable. Apart from that, the forecast indicated good weather and the fine starlit night which is usual over the desert. This clear sky would give a perfect setting for astro-navigation, which was particularly important on that route, owing to the scarcity of radio aids in that vast wilderness of sand.

At the cruising altitude of 12,000 feet, the captain levelled off and, on checking his course, noticed that there was a 25° difference between the readings of his magnetic compass and the gyro compass.

An aircraft magnetic compass is in essence a simple magnet encased in alcohol to dampen its oscillations. It is a remarkably reliable instrument, with two big weaknesses. In a turn or during a period of acceleration or deceleration, the liquid is not sufficient to stop the needle swinging widely from side to side. And if any metallic material is introduced into the cockpit or there are any changes in electrical current near it, considerable errors may result. For these reasons, all modern aircraft have a gyro compass, with the master unit far away from the cockpit in an area of magnetic stability. The needle on the gyro compass remains quite steady during any manoeuvre.

The captain reported the discrepancy between the compasses to the navigator and asked him to check the true course with the astro-compass. This is an instrument which uses the stars, not a magnet, to find direction, in the same way as we can tell north by finding the north star. The officer did this, and reported the gyro compass correct and the magnetic compass in error. The captain had checked the electrical circuits and could find nothing that would cause deviation. He should have returned to Tripoli, but continued the flight in breach of his company's regulations. No attempt was made to check the track made good by radio bearings, but the navigator kept up a series of star fixes as a check on his dead-reckoning navigation, while the captain continued to steer on his gyro compass, ignoring his magnetic compass.

At 01.24, the position according to the navigator's star fixes was twelve miles east of the direct track and a little over half-way to Kano. The difference between the compasses had increased to 54°.

At 03.24, a two-star fix indicated that the aircraft was a hundred miles west of Kano and, as the estimated time of arrival was in about half an hour, the navigator tried to tune in to the radio beacon, but without success. Meanwhile, the radio officer had received a meteorological report from Kano that it was experiencing electrical storms – which was strange, as there was no sign of any such disturbances in the clear night ahead of the Hermes.

Around 04.00 hours, when the aircraft should have been over Kano, the engineer officer noticed that the magnetic variation setting control on the gyro compass master unit was set to 60° west. The earth's magnetic field varies from place to place over the globe, and this makes the compass needle point a number of degrees either east or west of north. Magnetic variation is plotted accurately on all aircraft maps, and in navigation must be *subtracted* (for westerly variation) or *added* (for easterly variation) to obtain a true course. Over the area in which the Hermes was at that time flying, the variation was 6° west.

The variation setting control enables the variation to be incorporated in the compass reading, so that a *true* course can be steered, if such is required. The navigator had misinterpreted the graduation, and had initially set 30° (instead of 3°), increasing the variation progressively to 60° (instead of 6°) as the flight progressed. When the variation setting control was turned to zero, both the magnetic compass and the gyro compass agreed.

Nothing was the matter with the magnetic compass and, as a result of ignoring it, the captain had steered a series of courses between 27° and 54° in error.

While the navigator imagined he was near Kano, he was in fact, hundreds of miles out in the desert. Throughout the six-hour flight he had been taking star fixes. These should have shown immediately that he was miles away from his dead-reckoning position. In fact, all his star fixes faithfully followed his dead-reckoning.

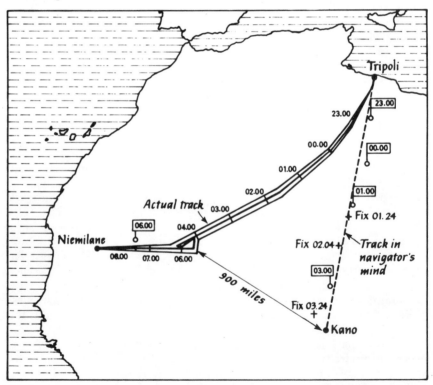

How did that come about?

In astro-navigation, you start work from an approximate dead-reckoning position, and with the periscopic sextant (due to its limited field of vision) azimuths and altitudes of a given star are calculated so that there is no difficulty lining up on pre-computed settings, That night the navigator had imagined the right star to fix his preconceived position.

As a result, as can be seen from the map, his astro fixes and his dead-reckoning positions followed each other closely.

When the situation was reported to the captain, there was considerable confusion. An attempt was made to work backwards to try to find out an accurate position, but while this was going on, no decisive action was taken. Reconstruction of the flight showed that the aircraft was about 900 miles north-west of Kano, but the crew thought they were further north, and set course for Port Etienne.

At 05.32 emergency procedure was adopted, and an SOS was sent at 05.58. The aircraft was then in the middle of the Sahara, and was somewhat short of fuel.

At 07.22 the Hermes managed to contact Dakar and thereafter received true bearings regularly. But about an hour later, the crew realized they had insufficient fuel to reach the coast, and would have to crash-land in the desert.

At 08.45 the aircraft descended and, after circling a village, belly-landed at a point about seventy one miles south-south-east of Atar. A massive desert rescue was organized and managed to be a considerable success in spite of many difficulties. Only six passengers were slightly injured, but the first officer died five days later as a result of exhaustion brought about by strain and heat.

The navigator had used his imagination to keep his own little world of the Hermes straight and level and meaningful and on the track for Kano in his own mind. But this little world and everyone in it was in reality on a fantasy flight nowhere near where they thought they were.

How far is our world like the little world of the Hermes? How far fantasy, how far reality? Looking through his sextant up at the stars that night, what did the navigator *really* see?

4

Not in Our
Stars But in Ourselves

'The girl who walked alone' – that was what the stars indicated about Amelia Earhart and that was what she was known as at school. Introspective, a poetess, child of parents divorced because of her father's drunkenness, not pretty but with a chiselled determined face, living in a small cell of privacy – clearly her future lay in the devoted social work with which she started her career. No one could have foretold that this girl would be a heroine known to millions and would become almost as popular as Lindbergh.

His solo flight across the Atlantic on 20–21 May 1927 had brought him a reported 3½ million letters, several thousand offers of marriage, 5000 poems, 1400 gift parcels and three invitations to go to the moon in a rocket.

It had also inspired a number of women to try to follow him across and become the first 'Lady Lindy'. Most of them never got further than talking about it. After all, the cost would be prohibitive. The women would have to be rich themselves, or sufficiently persuasive to interest a man to sponsor them.

A rich American woman born in Pittsburgh, Pennsylvania, was the first to enter the lists. She had married an Englishman called Frederick Guest, who had been in the Air Ministry under Lloyd George, and she wanted to be the first woman across the Atlantic. Seeing herself also as the ambassadress of good will between her country and that of her husband, she had purchased a three-engined Fokker – the same type as Kingsford Smith was to buy for his Australian National Airlines, except it had floats – from Commander Byrd, inventor of the bubble-sextant for navigation, who had himself flown the Atlantic, ending up by crashing on the French coast. Mrs Guest changed the plane's name from *America* to *Friendship* and began to make her plans.

Meanwhile other contenders were catching up. Chief of these was Princess Lowenstein-Wertheim, a British woman who had married a rich German prince who had been killed fighting for the Spaniards against the Americans in 1899. Sixty years old, she was a familiar figure in her leather coat and fur-lined hat, flying as passenger in open cockpitted aircraft along the European air routes. She also bought a Fokker triplane and engaged two experienced airmen, Minchin and Hamilton, to fly her 'the hard way' against the strong prevailing westerlies to Montreal.

The aerodrome at Upavon was chosen for the heavily laden take-off because it had the longest run over the grass. The three of them flew down there on 30 August 1927, and spent the night supposedly resting before their ordeal. However, the RAF organized a party to celebrate. Hamilton drank too much and spent most of the night playing melancholy music on the piano and prophesying disaster.

Next day, the Bishop of Cardiff christened the Fokker *St Raphael* after the patron saint of airmen, and they just managed to stagger off into the air, missing telephone wires at the far end of the field by inches.

The forecast was bad – strong winds and blustery weather. They carried no wireless. Half way across the Atlantic, their lights were spotted by an oil tanker. They were never seen again.

The next contender was another rich American woman, Mrs Frances Grayson. She set off with her pilot Stulz and navigator Goldsborough on 17 October 1927. Five hundred miles out, the port engine failed on the Sikorsky amphibian and Stulz returned. Mrs Grayson tried again on Christmas Eve with a new pilot called Omdal. The weather could not have been worse, with gales, snow, ice and hail forecast right across the ocean. A ship reported hearing the sound of engines, followed by a loud noise. Nothing more was ever heard.

Into Mrs Grayson's shoes stepped the Honourable Elsie Mackay, daughter of the rich Lord Inchcape, chairman of P. and O., the shipping line. A millionairess in her own right, she was also an actress, a horsewoman and a pilot. Like the Princess, she wanted to be first across and first east to west.

She hired an experienced pilot called Hinchliffe[1] at almost a thousand pounds a year and £10,000 insurance, and sent him to America to choose a suitable plane. He had had a distinguished war record, but had lost his left eye and always wore an eye-patch.

Hinchliffe arrived back with a single-engined Stinson which they called *Endeavour*. The whole enterprise had to be kept secret lest Elsie Mackay's father hear and stop them. On his part, Hinchliffe desperately wanted the money, since he had married a Dutch girl called Emilie and had two daughters. With only one eye and pilot medicals about to be introduced, he would shortly lose his one means of supporting them.

The trip was to start from Cranwell, the RAF College near Grantham, where Hinchliffe, his wife and a pilot friend called Sinclair stayed at the George Hotel. Elsie Mackay arrived later with two maids and two chauffeurs.

Lord Inchcape heard of the trip and sent his brother to the George to try to stop her. He was fobbed off – as were the press – with the story that it was Sinclair who would accompany Hinchliffe, not Elsie Mackay.

Endeavour had overstayed its welcome at Cranwell while they waited for good weather and the RAF were demanding that they move it. A further pressure was the announcement that a German crew headed by Baron von Huenefeld were making preparations to fly east-west from Dublin.

In spite of the fact that Hinchliffe was superstitious, preparations were made to take off on 13 March 1928. *Pâté de foie gras* and chicken sandwiches were made, thermoses filled with coffee. The charade that Sinclair was going was

1 See 'Beyond the Inquiry'.

kept up for the masses of pressmen waiting at Cranwell. Only at the last minute did a figure in a leather coat and helmet, muffled to the eyes, take his place.

The weather forecast was terrible. Over snow and slush, the black and gold *Endeavour*, laden with enough fuel for forty hours flying, pounded a full mile before lifting off into the overcast.

At 1.10 pm the aircraft was sighted in south-west Ireland, having completed only 400 miles in five hours. Ships in the Atlantic reported westerly gales and mountainous seas, but none reported the Stinson, which never reached the other side.

Elsie Mackay and Hinchliffe had vanished.

Young Emilie Hinchliffe was left quite alone with two small daughters. She now learned that Lord Inchcape, as trustee of his daughter's estate, had frozen all her assets. Not only was the salary due to Hinchliffe not to be paid, but the insurance policy was void because the premium had not been paid.

A strange series of events began eighteen days later. An elderly woman with some psychic powers who had consulted the medium Eileen Garrett about her son, killed in the First World War, came to see if further messages might come through her ouija board.

But the pencil wrote, '*Can you help a man who was drowned?*' and afterwards '*I was drowned with Elsie Mackay.*'

Further messages from the spirit of Hinchliffe urged her to contact his wife, giving the name and address of a Croydon firm of solicitors. She wrote to Sir Arthur Conan Doyle, whom she knew as a prominent spiritualist, enclosing a transcript of the messages. She also wrote to Mrs Hinchliffe.

Conan Doyle arranged for her to have a session with Mrs Garrett, and had the transcript sent to him. Amongst the messages was the information that Hinchliffe had turned south, that his wife was not English and had two small children. Conan Doyle was sufficiently impressed to write to Mrs Hinchliffe, and it was eventually arranged that she should see Eileen Garrett.

The session produced enough evidence of personal things to convince Emilie Hinchliffe. The spirit of her husband described the crippling of the Stinson in a storm and his turning away from the teeth of the head winds to try to reach the Azores. This in such circumstances would have been the logical thing for an airman to try to do, though it is difficult to see how any navigation was possible.

The spirit went on to say that there was no death, but an everlasting life. 'Life here is but a journey and a change to different conditions.'

A further session described his flight and his attempts to go north and the crack up in the storm before turning south and landing in the water near the island of Corvo in the Azores group. 'Currents may reveal wreckage' was one of the messages.

Other sessions with Mrs Garrett reassured her that she would receive the insurance money in July. This appeared unlikely as Lord Inchcape had given the money – amounting to several millions, invested and in trust – to relieve the national debt, and the Treasury would not part with it. Yet on 31 July Winston Churchill, the Chancellor of the Exchequer, rose to inform the House that £10,000 would be released to clear financial obligations appertaining to the gift, and this very considerable sum was given to Emilie Hinchliffe.

Almost a year after the tragedy Emilie Hinchliffe received a letter from the Air Ministry:

Dear Mrs Hinchliffe,
I am sorry to trouble you with past history, but I think you should know that the Air Ministry received a report in December last, to the effect that part of an undercarriage of an aeroplane was washed ashore in County Donegal bearing the following numbers and description:

76168547 Goodrich Silvertown Cord Airplane
150/508. Manufactured by the Goodrich
Company, Akron, Ohio, U.S.A.

As a result of a subsequent inquiry through the American Embassy, a letter was received from the Goodrich Rubber Co., to the effect that the undercarriage in question was part of the aircraft used by the Hon. Elsie Mackay and Captain Hinchliffe on their Atlantic flight last year.

The easterly flowing Gulf Stream curves round the Azores. Certainly this wreckage from the Stinson had been brought by currents. But there never was any further sign of Elsie Mackay. Another lady had vanished.

Princess Lowenstein, Mrs Frances Grayson, the Honourable Elsie Mackay . . . after weighing up the chances of success, Mrs Frederick Guest now decided to look out for 'a real American girl with the right image' to take her place in the Fokker across the Atlantic.

George Palmer Putnam, publisher, was looking for that same person. He had persuaded Lindbergh to write the story of his flight and had made a considerable fortune. Now he was searching for a 'Lady Lindy' to double it. Ideally she would need to be an innocent young female, with guts, a pilot's licence and no idea whatever of the real money value of any book she managed to write if she just happened to fly the Atlantic and stay alive.

Mrs Guest's and George Putnam's search led them to a refuge called Denison House in Boston, where Amelia Earhart worked mainly among women deserted by alcoholic husbands.

Amelia Earhart was the daughter of a lawyer and the granddaughter of a rich judge in whose big house overlooking the Missouri she was born on 24 July 1898 – missing by a whisker coming under Cancer the Crab, but having every bit as much determination to hang on.

Her grandfather never thought much of her father and put such pressure on him that he took to drink and became an alcoholic, lost his job and separated from his wife. As a result Amelia Earhart, at a very early age, learned three things that influenced her life: how not to be dependent on anyone, an understanding of alcoholics (perhaps because her father was the only man she loved) and a missionary zeal to improve the world for women.

During the Great War, she nursed Canadian wounded in Toronto, studied medicine at Columbia University, and became interested in aviation from a trial flight her father arranged for her. She soloed in 1921.

Being a pilot with total freedom and independence in the air fitted her needs like a glove. Wearing khaki trousers, knee-length boots, and a purposely greased and dirtied leather flying jacket with helmet and goggles, she was the epitome of the new liberated woman – but in the nicest sense, and in spite of the get-up, she still looked a girl.

In fact, she probably should never have been a pilot at all. Though Marion Perkins, head worker at Denison House, Boston described her as 'an unusual mixture of the artist and the practical person', Amelia was an introspective artistic loner. She had no inborn piloting ability – her landings were poor and she was to crash four times. Elinor Smith, who was a born pilot, went up with her and felt awkward at her ineptness. The performance did not make any sense, she said, 'except for Lady Heath's speculations that it was always Earhart's co-pilots, (or "mechanics", as Putnam called them) who had done her flying'.

Amelia wrote poems that largely showed her motives and character:

> *Courage is the price that life exacts for*
>   *granting peace.*
> *The soul that knows it not knows no release*
>   *from little things.*
> *Knows not the livid loneliness of fear,*
> *Nor mountain heights, where bitter joy can hear*
> *The sound of wings.*

Her father encouraged her and, with his help, in 1922 she bought a biplane and built up her flying hours. Two years later her parents were divorced and, disappointed and discouraged, she sold the plane and went to live with her mother and sister near Boston. There she eventually took up a position as a social worker at Denison House, where George Putnam found her.

She was perfect! With her short hair and good-looking face, she even looked like Lindbergh – and she wrote, too. Putnam wasted no time in passing the good news to Mrs Guest. Amelia Earhart accepted the position of understudy and she was flown up to Trepassey Bay, Newfoundland, in the seaplane *Friendship* by Stulz (the ill-fated Mrs Grayson's ex-pilot) and mechanic Gordon.

Trepassey Bay was the harbour from which the three navy Curtis flying boats had taken off on 16 May 1919 to fly the Atlantic via the Azores over a bridge of forty-five American warships. Only one succeeded. Amelia Earhart was disappointed that her role on board was to be just a passenger and apprehensive to discover that Stulz was drinking a bottle of brandy a day because he was scared. Alcoholism was to dog the teetotal Amelia as later it did Amy Johnson, but at least she was an expert in how to cope with it. On 17 June 1928 *Friendship* left Trepassey Bay and after a stormy passage landed at Barry Port, Wales.

Putnam had been quite right about the publicity bomb going off, just as it had done under Lindbergh. There were so many offers for personal appearances, so many opportunities for an ever-widening golden future that she was overwhelmed. It was natural that George Putnam, the publicity expert, should help and guide her. He sat her down in his own house to write *20 Hrs, 40 Mins*, the story of her Atlantic flight. By this time Putnam realized what he had hit, and resolved to take this goldmine to be his lawful wedded wife.

With his helmet of black hair and rimless glasses, twelve years older than she was, Putnam was no pin-up. Though rarely liked, he could put on the charm and he was possessed of ruthless determination. The fact that he already had a wife and two children in no way deterred him from his purpose. The degree of Amelia Earhart's innocence can be seen from the fact that she dedicated her book to his wife.

Putnam dedicated himself to the promotion of Amelia Earhart. He introduced her to film stars, writers and politicians. He entered her for the first women's air race, the Powder Puff Derby, in which she came third. He organized all the details for her breaking the women's speed record at 181 mph. Now divorced, he also began regularly proposing to her.

On 7 February 1931 she at least agreed to marry him, against her mother's wishes. It was a business arrangement, probably brought about by her sense of loss at her father's death. Before the honeymoon she presented him with a cold note, stipulating they should both be at liberty to go their own ways.

Her way was up and up – with more and more publicity as her husband capitalized on her name in every conceivable way. Record after record followed. Then on 20 May 1932, the fifth anniversary of Lindbergh's flight, she set off in a Lockheed Vega to fly alone across the Atlantic. The altimeter went wrong. The exhaust manifold burned, allowing naked flames to play on the wings and fuselage. She flew through storms and ice. But she did reach the other side, landing near Londonderry – bending an undercarriage strut at the same time – in fourteen hours and fifty-six minutes.

More and more honours showered down on her. She won a $10,000 purse for the being the first person – man or woman – to fly solo the 2400 miles from Hawaii to California.

Like a number of airmen and airwomen, she became deeply interested in extra-sensory perception, following with great excitement the work of Dr Rhine of Duke University.

Perhaps the most satisfying role to her was that of counsellor to women students at Pardoe University. There she preached the need for women to learn independence and the ability to stand on their own two feet. 'If we begin to think and respond as capable human beings, able to deal with and even enjoy the challenges of life, then we surely will have something more to contribute to marriage than our bodies. We must learn true respect and equal rights from men by accepting responsibility.'

One thing led to another. Pardoe University, delighted with her work, voted $50,000 for her to buy a plane to fly round the world.

The aircraft she chose was the latest in the world, a ten-passenger, twin-engined, all metal aeroplane from Lockheed. It was not an easy aircraft to fly. The aircraft type was called the Electra after the missing star of the Pleiades.

To go with her, Amelia chose Fred Noonan, one of the most experienced navigators in the world, who had navigated the American's first commercial service from California to Hawaii in a Martin flying boat two years before, but he was also a chronic alcoholic. She also had with her another experienced navigator, Captain Manning, since two navigators were considered necessary because of the danger of flying the vast Pacific.

They set off on 17 March 1937, westwards from Oakland and reached Honolulu without incident.

There they loaded up with maximum fuel. Amelia opened up the throttles.

The nose swung first to the left, then to the right. Juggling with the throttles to try to keep straight, Amelia got into a ground loop, swinging round in a complete circle. The undercarriage collapsed and one wing was torn off. The crash was tactfully put down by the US Army to 'unexplained circumstances'.

The aircraft was shipped back to California. While they waited for it to be repaired Manning excused himself to go back to his ship. Fred Noonan used the time to get married. He was determined to go on with the flight, needing the publicity to start a navigation school to support his wife.

There was the same sort of missionary relationship – indeed they looked not dissimilar – between Amelia Earhart and Fred Noonan as between Katherine Hepburn and Humphrey Bogart in *The African Queen*.

Noonan promised to kick the alcohol habit but on the honeymoon he drove up the wrong lane head-on into another car. The police ticket read: '*No injuries. Driver had been drinking.*'

On 20 May 1937 – anniversary of both Lindbergh's and Amelia's Atlantic flights – she and Noonan set off again from Oakland, California in the repaired Electra on their round-the-world flight, this time flying in the opposite direction from west to east. They flew to New Orleans, Miami, Puerto Rico, Brazil, across to West Africa, Mali, Chad, Sudan, Karachi, Calcutta, Akyab, Singapore, Timor, Port Darwin, finally reaching Lae, New Guinea on 30 June. Amelia wrote despatches on this flight for publication by the *New York Herald Tribune*.

The next flight was the most dangerous – 2556 miles over open sea to Howland Island, two miles long and half a mile wide. The USS *Ontario* was stationed in the middle and the cutter *Itasca* was anchored off the island to act as a radio centre. The aircraft had been lightened of every possible item including the two parachutes. For some extraordinary reason (Amelia said it was too much trouble to reel in) the 250-foot trailing aerial for the radio was removed.

At 10.30 am on 2 July Amelia opened up the throttles and set off on an easterly course on an eighteen-hour flight plan. They were in fact flying into yesterday because they would be crossing the international date line, and their take-off time Howland was 12.30 am on 1 July.

At 07.20 GMT they sent a position report, latitude 4°33' south, 159°06' east, 795 miles east of Lae on course groundspeed of 111 knots.

At 17.45 GMT, fifteen minutes before the estimated time of arrival, Amelia reported, '200 miles out and no landfall,' and whistled too briefly into the microphone for a bearing.

At 18.16 she called, 'Approximately 100 miles from *Itasca*. Position doubtful.'

At 19.28 she reported , 'Circling trying to pick up the island.'

Three naval operators at Honolulu received a message saying, '*281 North Howland, call KHAQQ beyond North, don't hold with us much longer. Above water, Shut off.*' It was in code – which was strange for an emergency – and was assumed to mean that Amelia Earhart had ditched 281 miles north of Howland.

The *Itasca* was sent to search the area. Numerous other reports were received from professional and amateur wireless operators, including one from Pan American Airways who had wirelessed Amelia Earhart to come in on 3105 kilocycles using not a voice signal but dashes – and immediately afterwards four dashes were heard. The message came from around 150 miles south of Howland, an area so far not searched.

Other signals seemed to indicate that Amelia Earhart was still transmitting – which could only be possible on land using one of the engines to generate power. George Putnam thought they had landed at the Phoenix Islands, 280 miles south of Howland, and appealed to the United States Navy to search there.

As in other disappearances, almost everyone seemed to have ideas on where they had come down, and 'messages' came flooding in long after their petrol must have been exhausted.

On Roosevelt's orders, the US Navy mounted a $4 million search with the aircraft carrier *Lexington*, the battleship *Colorado* and a whole flotilla of other warships.

Nothing at all was found, and she and Noonan were declared 'lost at sea'. George Putnam inherited Amelia's estate, sold his publishing business, married again twice and died of uremia poisoning.

The theories of what happened to Earhart and Noonan multiplied with the years and still remain a mystery.

One theory was that they were lovers who simply disappeared together (like Mrs Miller and Bill Lancaster[1]) to live happy ever afterwards on a Pacific island – quite out of character.

Another theory was that instead of flying north-east towards Howland, they flew due north to Saipan, then a secret Japanese base, where they sent radio messages before being caught and executed. Several witnesses reported seeing two American fliers, one a woman with short hair, on the island. Two bodies found there were flown to America for examination, together with part of an engine found in the harbour believed to be from the Electra. How the radio messages could have materialized was not explained, and anyway the 'clues' proved negative.

Another theory put forward by a radio journalist called Fred Goerner, who explored the area for six years and wrote a fascinating book[2], was that Amelia and Fred were on a spy mission for the American government to Truk, also a secret Japanese base, taking a big dog-leg north of their course, before proceeding towards Howland.

A runway had been constructed on the island, according to Goerner, ostensibly to accommodate Amelia Earhart's Electra, really to have available for land-plane use to spy on what the Japanese were up to in the Marshall Islands. Pan American commercial services only used flying boats, and the American government did not want to offend the Japanese by overt military activities.

Goerner believes that they flew so far north on their sun line that they reached Mili Atoll in the south-eastern Marshalls, where Noonan was hurt in a crash-landing. Amelia sent out SOS on the emergency radio transmitter

---

1 See 'The Other Side of Luck'.
2 *The Search for Amelia Earhart*, Bodley Head, 1966.

which was picked up by both Japanese and American ships, and a race was on to get to them. The Japanese won, taking them to Saipan where, according to some witnesses, Amelia Earhart died of dysentery and Fred Noonan was beheaded.

But if they were on a spy mission for the government, why were preparations so offhand, why only one navigator (and an alcoholic at that), why no proper long-range radio, and why not simply a return to Lae after 'spying' with the perfect excuse of bad weather?

There remains the simplest explanation. Amelia Earhart, like all those early fliers, knew the dangers. Echoing the wish of many of them, she had written, 'When I go, I'd like best to go in my plane. Quickly.'

A rather strange incident had occurred on the trip across the South Atlantic flying from Natal in Brazil to Dakar on the West African coast. Over the ocean no sextant shot had been possible due to heavy overcast, and when the African coast was sighted, Noonan insisted they should turn right to the south. Had they done so, they would have reached Dakar, eighty miles away. But a 'left turn seemed to me to be in order', said Amelia in her diary, and after fifty minutes flying found themselves over St Louis, Senegal, 163 miles off course.

Noonan passed up a note. 'What put us north?'

Amelia must simply on instinct have made a turn 180° different from the one given to her by her navigator, and not even told him! Any self-respecting navigator would have off-loaded himself as soon as they reached the ground – especially as ahead of them lay the Howland leg. And though they had radio telephony, in those days its range was minute and its performance erratic, and they had no long-range morse radio.

Flying to a tiny island over hundreds of miles of ocean was the most dangerous aviation experience possible in those days. Astro-navigation was not so advanced and getting a fix on stars or sun and moon (which was all Noonan had) was very much open to error. There was no radar and radio direction-finding was still in its infancy. Twelve years later, when we used to fly between the Azores and Bermuda into the so-called Bermuda Triangle, we were always on the alert. Three British aircraft had disappeared on that route. Unforecast winds can suddenly spring up and radio conditions can deteriorate. Bermuda was very easy to miss. If ever an airman gets into a position of not knowing whether he has passed an island and is over on the other side, he is in serious trouble. What does he do? Go back? Go forward? Go left? Go right?

I have been in that situation only once, and had to radio for help. The Air Sea Rescue airborne lifeboat flew out from Bermuda and homed on my fully laden Constellation, keeping us company till we safely reached the island with just enough fuel.

Howland was six times easier to miss than Bermuda – and there was no airborne lifeboat to call. Another factor, not then discovered, was that we have time clocks inside us, regulated to the time of our normal environment. We all know now of jet lag, and that businessmen and politicians are warned not to make decisions too soon after flying for long distances in an east or west direction. But Amelia Earhart and Noonan had flown from 120° east to 180° west. As well as being tired and under strain, they would be twenty hours out in their normal time-clocks.

In a difficult situation, it is unlikely they would have been able to think straight. Amelia tried to get a bearing at 1000 feet. She must have known that she would have to climb at least to 4000 feet to extend the radio range adequately. She would have known to transmit longer so the *Itasca* could have a chance to get a bearing. Circling first and then running up a position line is not good practice when you are lost.

It is usually at the end of a long flight that airmen make mistakes. It was at the end of a tiring ocean flight that Amelia Earhart suddenly turned left instead of right – she might indeed have had a laterality problem that often lies latent in many people, turning left not right only under stress – an occurrence that has probably been the main cause of at least nine accidents in post-war civil flying. It is interesting that George Putnam pressed the Navy to search 280 miles *south* and not 281 miles *north* as had come over on Amelia's signal. When such time-clock discrepancies are added to fatigue, navigation errors are easy to make on a difficult and dangerous leg with inadequate radio and navigation equipment.

But then also look at the photograph of the positions in the aircraft of Amelia Earhart and her navigator. They were totally separated. Amelia has her private little room with the view. Noonan has the room at the back. Communication between them would have been extremely difficult. And the only way Amelia Earhart had of checking Noonan's navigation was to study it upside down!

Add tiredness, add jet-lag, add a possible left-right tendency, add a belief in her own sense of direction to that extraordinary separation between pilot and navigator. Mix in the vast Pacific and a tiny island, and perhaps you have the answer to the mystery.

Amelia Earhart was neither a born aircraft designer nor a born pilot. She was born a clever and courageous woman, determined to break through the barriers into a hitherto all-male world, who found in flying the ability to transcend the smallnesses of life and in the sky could become once again 'the girl who walked alone'.

In her last letter to her husband, she wrote: 'Please know I am quite aware of the hazards. I want to do it because I want to do it. Women must try to do things as men have tried. When they fail, the failure must be but a challenge to others . . .'

Three and a half years later, on Friday, 3 January 1941, a woman pilot who idolized Amelia Earhart and was a member of the Air Transport Auxiliary (an organization that ferried aircraft for the RAF) was instructed by Pauline Gower, the officer in charge, to fly an Oxford aircraft from Hatfield to Prestwick. There she was to collect another Oxford and deliver it to Kidlington near the city of Oxford.

The woman's name was Amy Johnson, heroine of the song, *Amy, Wonderful Amy*, the first woman to fly to Australia, record-breaker of the routes to Tokyo, to South Africa, across to America, where she had met and become the friend of Amelia Earhart, and a survivor of three unhappy love affairs.

On 3 January 1941 a long tongue of high pressure stretched from Scandinavia all over the British Isles. The weather was the usual winter anticyclone condition: dry, very cold, light wind, radiation and smoke fog round towns improving during the day. The ground was ice-hard and mostly covered in snow, but the cloud was scattered and high, and visibility was good.

At South Cerney Service Flying Training School, where I was learning to fly twin-engined Oxfords, flying continued throughout the day. I did two sessions of practice instrument flying with a fellow pupil in the area between the airfield and Oxford, thirty miles away.

The weather report Amy Johnson would have studied showed Renfrew as typically fogged in with Glasgow smoke. But Prestwick, on the coast thirty miles to the south, was well away from industrial haze, and had probably the best record for good weather of any British airfield. There was very little cloud over the Irish Sea en route – Eskdale a few miles right of track reporting a visibility of over thirty-one miles and a trace of cloud at 4000 feet.

She set off, but after only a hundred miles flying put down at Ternhill, southwest of Market Drayton, miles off her track. She night-stopped at the Hawkstone Park Hotel, Weston-under-Redcastle. Why? There was no radio on board so she could not have received a warning of deteriorating weather. The weather was the same as when she started and remained so throughout that night, improving all the time with the barometer rapidly rising. Apart from its excellent weather record, Prestwick was one of the easiest aerodromes to get into, particularly in conditions of light wind. There were masses of aerodromes on her route. Why didn't she continue?

If by any chance she could not get into Prestwick she would have loads of fuel to divert. If she was going to put down again so soon, why did she ever set out in the first place?

The weather next day, 4 January, was good for winter. Cloud base was generally around 8000 feet and visibility was between $2\frac{1}{2}$ and $6\frac{1}{4}$ miles. I flew four times – the longest flight an instrument-flying cross-country to Peterborough. En route, I passed the huge bomber station of Upper Heyford near Oxford, about twelve miles to starboard, with the smaller airfield of Kidlington just south of it.

Having reached Prestwick that same day, Amy Johnson was instructed to ferry another Oxford, V3540, to this same station of Kidlington.

She set off at 4 pm, at the same time as many training Oxfords were flying in good visibility and weather conditions in the vicinity of her destination. Her trip would have taken about an hour and forty-five minutes.

From the weather map and the actual observations at 01.00, 07.00, 13.00 and 18.00 GMT on that day, though it was still cold, the weather must have been reasonably good. What little cloud there was must have been between 3000 and 4000 feet, and the visibility around $12\frac{1}{2}$ miles, except for occasional patches of sea mist and fog.

Kidlington was so close to Upper Heyford that it would have the same weather, which throughout that day was: 01.00, hazy, $6\frac{1}{4}$ miles visibility and cloud base 4000 feet; 07.00, almost the same; 13.00, overcast, visibility $12\frac{1}{2}$ miles and cloud base 3200 feet; 18.00, cloudy, visibility $12\frac{1}{2}$ miles and cloud base 2500 feet.

There was nothing particularly difficult about that weather. Yet that evening, Amy Johnson's sister, who had been spending the afternoon with her husband, returned to her house in Blackpool to find Amy there, saying she had had a rather sticky passage from Prestwick and had fortunately been able to reach Squiresgate. Amy spent that night with her sister, going back to her Oxford at Squiresgate next morning.

On 5 January there was still no cloud in that area, but it had become particularly misty. En route, there was more cloud but the base generally still remained around 2500 feet. Kidlington weather at 07.00 GMT was 12½ miles visibility, and cloud base 2000 feet: at 13.00, 2½ miles visibility and cloud base 2500 feet.

Amy got into the Oxford and waited for the mist to clear, chatting with the refueller and smoking (not strictly allowed on RAF aircraft). To people who tried to dissuade her from going, she was reported as saying that she could smell her way to Kidlington and that she 'would go over the top'.

The first she could certainly have done under those reported weather conditions. The second was interpreted as meaning she could go above the cloud – and bearing in mind the Oxford had no wireless, that would have been lunatic. More likely she meant she would go over the top of the mist, vertically down through which the ground would have been visible.

But why hadn't she continued to Kidlington the previous day? Had she simply stopped at Blackpool to visit her sister? Nothing unusual about that – it was an unimportant flight and it was fairly normal for pilots to drop in on friends, give lifts, fetch drinks and food for parties (there was always an influx of RAF aircraft 'on training' into Northern Ireland for turkeys at Christmas time), to which the authorities turned a blind eye.

What is significant is that she had declared she could 'not now fly on a Sunday. All my crashes have taken place on Sunday'. Certainly she delayed one of her record flights so as not to fly on a Sunday. Why then appear to arrange things so that she *did* fly on Sunday?

The verbal reports of the weather that day were worse than the actuals, though certainly there must have been pockets of fog and hill cloud on that icy snow-covered countryside. Count Zichy flew a Hurricane from Henlow to Aston Down that morning, hugging the ground through bad conditions and ending up ground-looping within inches of a petrol tanker. He had seen Amy Johnson that Christmas, and to him she did not seem happy, waiting to go on a conversion course. Since she could never really bring an aircraft smoothly down to earth – her landings were once described as 'controlled crashes' – conversion to a new aircraft type would have been a strain. On the other hand, her sister described Amy as looking happier that morning.

She eventually opened the throttles at 11.49 and was immediately swallowed up into the mist.

Her movements since Friday had already posed a number of mysteries: why had she landed at Ternhill and stayed overnight at the Hawkstone Park Hotel? Did she meet somebody there? Why didn't she press on next day to Kidlington where weather observations show the weather as good? Why did she flout her own rule of never on Sunday?

Now an even bigger mystery began to materialize. At the time she flew diagonally across the misty country from the coast to the Midlands, weather observations record no cloud at all or a cloud base of around 2500 feet and a visibility of between $2\frac{1}{2}$ and $6\frac{1}{4}$ miles. Under such conditions, there should have been no difficulty in completing the trip below all cloud in an hour.

Yet from the moment she took off from Squiresgate nothing was seen or heard of her till three and three-quarter hours later when at 3.30 pm she was seen parachuting down from low altitude into the Thames Estuary, followed by Oxford V3540.

There was nothing at all mysterious in the background of Amy Johnson. The daughter of a teetotal Methodist herring merchant in Hull, her main claims to fame at school were that she was the only girl who bowled overarm at cricket and that as a teenager she led a revolt against the wearing of uniform straw hats. She took an ordinary degree in economics, French and Latin, and became a typist.

Then in 1924 she met a Swiss businessman working in Hull. And two years later she had her first flight in an aeroplane. She fell in love with the former and was unimpressed with her 5/– flight – 'it was all over so quickly, nothing happened'. She became the man's mistress and would have also embraced his religion, but her parents were horrified, and she was banished to London where she became a shopgirl at Peter Jones.

Amy's interest in flying appeared to have died before it even began. All her passion seemed centred on her boy-friend, now in America, to whom she wrote several hundred letters.

Then he wrote to say that he was taking a girl out to theatres and dancing. She retaliated by going out with other men and by joining the London Aeroplane Club. Flying was a still daredevil thing – especially for women – and clearly she hoped to give him a shock.

It was she who got the shock. In the summer of 1928 he came suddenly to London, took her out to dinner and told her that his companion for theatres and dances was now his wife.

Amy was so upset that she thought of suicide, and began considering flying as a means to that end, throwing all her energies into it, working voluntarily at the Air League and meeting such intrepid women solo fliers as Lady Heath and Lady Bailey. And on 15 September 1928, helmeted and goggled, she had her first flying lesson. At the end of half an hour, her instructor told her bluntly that she was not a pilot.

That was the fuse that sent her on her way. Unlike Amelia Earhart, who wanted to show men what women could do, Amy Johnson wanted to show *one* man what she could do. Suicide was forgotten in the exhileration of struggling towards that impossible goal.

Her instructor had spoken the truth. She had little co-ordination. She could not judge height above the ground – an indispensable flying skill that simply cannot be taught. Years later, the Schneider Trophy pilot Atcherley gave her instruction on instrument flying, and was shocked to find how incapable she was of flying blind.

But she had two qualities – guts and determination. Powered by an abundance of both, she rocketted to fame as – of all things – a pilot.

After an inordinate amount of dual instruction, she managed to go solo. She then took her A and C engineer's licences, the first British woman ever to do so.

An *Evening News* reporter asked her plans, and she told him she wanted to make a long distance flight. *Girl to Fly Alone to Australia* was plastered over the front page.

And the girl who flew 'like a good engineer' was practically on her way. In fact she would never have done it if she had not been an engineer. Her father contributed £500 to buying her a plane. Lord Wakefield, the oil magnate, underwrote the other expenses. She passed her B pilot's licence and bought a Gipsy Moth – almost identical to the Tiger Moth on which thousands of World War II pilots had their initial training, and an easy aeroplane to fly.

She called it *Jasmin*, a contraction of her name and, according to her father, 'lucky'. That year she was wearing green, so it was painted green. 'I should like green, please,' she wrote to a firm anxious to contribute an alarm clock for her trip. 'My machine is green, and also everything else I could get to match, as green is my 'lucky' colour.'

Green in fact is usually considered an unlucky colour. Samuel Franklin Cody, who first flew on British soil, would not have the colour at any price anywhere. Women coming to his house in a green dress were hastened upstairs by his wife to change into some other colour. Not a speck of green was allowed on his aeroplanes nor on anyone who flew in them. It was ironic that in the seaplane crash in which he died, his passenger, who was also killed, was wearing green socks.

Certainly green seemed to bring Amy luck in *Jasmin* to Australia, since no other flying record seeker ever escaped from so many near disasters and yet won through.

She had a couple of hundred hours flying and hoped to beat Squadron Leader Hinkler's record of twelve days.

First attempt at take-off from Croydon nearly finished on the fence. Her undercarriage collapsed on landing at Baghdad. At Bander Abbas 'I landed fast, as usual, and rather heavily, also as usual,' and her port wing went. At Insera she crashed and the propeller, wing and undercarriage broke. The pieces were glued together and the fabric repaired with men's shirts. Fortunately she carried a spare prop. At Java she ripped her wings, repairing them with sticking plaster. Landing at Timor, she had to weave in and out of ant-hills six feet high, amazingly hitting none of them. Taking of she touched the tree tops. She finally reached Australia, though it took twenty days.

She was world famous overnight. Even a final and almost total crack up, overshooting at Brisbane during a round-Australia flight, failed to damp the furore.

Amy was flown from Brisbane in *Southern Sun*, destined to take part in the search for *Southern Cloud*[1] ten months later and also crash. Ulm, destined to be lost over the Pacific, was the captain. Jim Mollison, destined to be a record-breaking pilot, her husband, and (in her own words) 'my undoing', was the first officer. Used to being idolized by glamorous women, he was never in love with her. In business ways a replica of George Putnam, he saw the affair as an excellent publicity arrangement – two famous pilots married to each other, the most tremendous box office.

1 See 'The Other Side of Luck'.

She next met Mollison in Capetown, meeting him at the airport when he came in after breaking the England–South Africa record. She herself meanwhile had flown to Tokyo in the second attempt – the first ended in a crash in a potato field – with Charles Humphreys.

Six weeks later, Mollison proposed at Quaglino's restaurant. She replied, 'I'll take a chance.'

Both world famous figures by now, they were married in a glare of publicity on 29 July 1932. They lived in luxury hotels. They mixed with high society. Mollison started affairs with other women, and drank not only on the ground but in the air.

He also had a psychological reason for flying. He flew to conquer fear.

Both driven by their own needs, they did marvellously well. Mollison completed the first solo east–west North Atlantic and South Atlantic flights. Amy beat his record to South Africa. Then the two of them on the second attempt – the first crashed – flew across the Atlantic, landing up with the aircraft on its back at Bridgeport, New York and Mollison unconscious.

They tried to fly to Australia together in an aircraft called *Black Magic*, burning out three pistons in a fast time to Allahabad.

After that failure, Mollison's drinking increased. Amy left him to go to Paris, meeting a rich sugar-daddy who helped finance her flying. Trying for a new England–South Africa record, she borrowed a plane, damaged it on her first attempt and broke the undercarriage on the second. Trying a third time in a new aircraft, she ground-looped it on take-off, damaging it badly. On the fourth attempt, she smashed the old record, also setting up a new record for South Africa–England and for the round trip.

There was one last futile attempt to patch up her marriage with a projected two-some round-the-world trip. Then she started divorce proceedings. Her French sugar-daddy pressed her to sleep with him. She refused, and he married a beautiful Hungarian girl. Amy came back to England – short of money, rather desperate and still in love with Jim Mollison.

It was natural that when the Second World War came, she should join Air Transport Auxiliary. It was a good steady job. It was flying. And it kept her occupied.

They put her on Ansons – the easiest twin-engined aircraft to fly. In comparison Oxfords which she also flew were real devils, particularly to land. Designed by Nevil Shute, the novelist, they were very sensitive on the controls and they swung, bounced and ground-looped.

From such an aircraft, on that cold January afternoon, parachuted the woman as popular as Our Gracie Fields, renowned world-wide as 'Ahr Amy who has more backbone than one of her father's kippers'.

An American reporter called Drew Middleton saw the parachutist first from the deck of one of the escorts of a small coastal convoy, dropping from the clouds at low altitude.

On the bridge of another escort, the motor launch *Haslemere*, Lieutenant O'Dea, second in command, reported, 'Suddenly I saw a parachutist come through the clouds. Shortly afterwards a plane came into view through the clouds near the parachutist. I did not hear any machine-gun fire.'

The *Haslemere* rushed towards the parachutist who had landed in the water half a mile away, only to run aground. While the ship was struggling free, the parachutist was swept towards the stern. A seaman heard the words, 'Hurry, please hurry,' and thought at first they were spoken by a boy.

Then he and another seaman realized it was a woman in the water, apparently wearing a life-jacket.

But both said there was someone else in the water. Commander Fletcher, Captain of the *Haslemere*, dived in to rescue this other person, while the seamen tried to reach the woman to drag her out. But the stern came up in the heavy swell and dropped on top of her.

A lifeboat was launched which followed the commander in the water. Men in it saw Fletcher reach the person, whom some seamen on the lifeboat described as wearing a helmet and a scarf. But whether alive or dead was not known, and only Fletcher was pulled into the lifeboat, already unconscious in the icy sea. He died later without recovering consciousness.

Another witness that day was Bruce Waugh:

I was an Ordinary Seaman on aircraft look-out on the bridge of the Hunt Class Destroyer *Berkeley*. We were escorting a convoy from Portsmouth to the Thames Estuary.

The weather was bitterly cold with strong winds and snow flurries and low cloud – in fact a truly miserable day. We were, as far as I now recall, approaching the Thames Estuary.

Through my binoculars I suddenly saw a plane on the port bow and immediately gave the alarm. As the plane drew nearer, I could see it was a yellow trainer and therefore identifiable as British and we then could all relax. But as the plane approached I could see it was losing height, though otherwise it seemed under control. But its descent continued until it finally crashed into the sea some 500 yards ahead of us.

The captain immediately ordered full speed and we raced towards the wreckage we could see floating on the waves. We approached to within about seventy-five yards and stopped engines while a boat was lowered to go to the scene. In about fifteen minutes, the boat was back alongside and reported that no bodies had been seen. They brought back an attaché case which, when opened, revealed the owner to be Amy Johnson.

Meanwhile an ML [Motor Launch] had also raced ahead and approached the wreckage. I saw a figure from the ML dive overboard (presumably because he had seen a body). He was pulled inboard a few minutes later and I subsequently learned that the exposure in those icy seas had killed him.

I had not seen any parachutes descending and my impression was that the pilot had deliberately approached the convoy in the hope that when the aircraft crashed they would be picked up.

Just as the ship got under way again a German plane swooped out of the cloud and dropped a bomb which landed just a few yards astern. Had the ship delayed a few more seconds in getting under way the German would have scored a direct hit. As it was, the percussion under water did some damage to the engine room which necessitated some days in dock for repair.

Amy Johnson has been positively identified as the parachutist. But was there another person in the water? If there was not, why did Commander Fletcher (who would have powerful binoculars) dive into the icy and dangerous water? If there was, who was he?

All the seamen of the *Haslemere* at the time and in the probate court held two years later swore that there was another person in the water 'beyond the shadow of a doubt'.

Only Amy Johnson was in the plane when she set off from Squiresgate – there were many witnesses to that. There was no record of her landing at a British airfield.

Could the 'body' have been one of Amy Johnson's two bags, both of which were picked up? Various newspapers declared it was. Others were certain of the existence of the mystery man.

Amy's body was never recovered, and the argument has gone on ever since. Various slanders inevitably crept in that Amy Johnson had been on a smuggling trip to France: that she had made a date with a lover and had lobbed down and picked him up somewhere.

One theory was that the man was a German airman who had been shot down, though there were no claims and reported losses that day.

Another theory was that she had been to France on Secret Service work and had picked someone up to bring to England. She had several times flown to France during the phoney war, and she had French connections. Three things support this theory: she was wearing a life-jacket which was not issued to ATA pilots because they did not fly over water: it would explain the 'absence' period mystery: and the weather over France that day was certainly better than in England.

The tactics of flying that short hop from Squiresgate to Kidlington were simple enough. There was fog but no cloud over the area where she took off. But fifty miles south-west at Anglesey the visibility was good and remained so all day. If things had got too bad, she could have turned back and diverted there.

En route through the industrial Midlands, visibility would have remained poor at around $1\frac{1}{4}$ miles, but the cloud base at 2500 feet was high enough. If the balloon barrages were up – not used in bad weather when there was no alert – they could easily be skirted. If the worst came to the worst, there was literally a stepping-stone line of aerodromes along her track, at any one of which she could have lobbed down.

The only reason why she should get lost in those weather conditions is that she panicked. But there was no evidence of panicking in her flying career. On the contrary, she coped with far greater difficulties than the weather that Sunday with the utmost resourcefulness.

Ice has been mentioned as a possible culprit – unlikely, I think, in those cold dry conditions. She had hot air to combat carburettor ice – and if there had been any trouble it would have manifested itself fairly soon in that uniform chunk of weather – and all the time would be below her, like the net under a wire-walker, those stepping-stones of aerodromes.

Airmanshipwise, therefore, that whole three and three-quarter hour trip and subsequent parachute descent was incomprehensible, unless she had diverted her track to pick someone up, perhaps in France.

Amy Johnson had approached Lord Vansittart at the Foreign Office and had asked for some really dangerous work, such as the Secret Service. An officially authorized trip is unlikely, but she may have arranged something through her contacts.

The weather does not appear bad enough for her to mill round lost over England for two and three-quarter hours. For one of the world's most press-on pilots, she had been remarkably dilatory delivering a couple of Oxfords. And there was something both naive and cloak-and-dagger about her whole personality anyway.

The theory could account for much of the mystery. She might have landed in France and picked someone up. The hurried descent from the skies might have been occasioned by an attack from enemy aircraft which, as Bruce Waugh's account demonstrates, were around.

But the most convincing evidence that there were two people in V3540 that day is embodied in the Oxford aircraft itself.

It was a flimsy machine made of wood which burned easily. It was also very touchy on the controls. One pilot saved his life when the Oxford was almost uncontrollable by steering between two adjacent buildings so that both wings were stripped off to burn themselves out harmlessly, while he careered on in the fuselage.

A ditching, particularly in those sea conditions, would have been suicidal. Quite apart from the plane immediately breaking up, the pilot would have had the utmost difficulty getting out. The roof exit, which was not directly over the pilot, was difficult to release and from the Oxford's beginnings there had been trouble with it. My instructor advised me not to try to use it. Certainly a parachutist would not have tried.

That left the door, way down at the other end of the fuselage. But even leaving the pilot's seat was appallingly difficult, since with a seat-type parachute you carried on your back not exactly your house but the cushion you sat on, and no man in full medieval armour waddled worse. Then the main spar went through the centre of the fuselage, covered with plywood, and up over this ridge you had to scramble before getting to the door.

Mindful that I might run into an emergency when I had to jump, on a solo flight I once practised trying to get out. The weather in which I did my experiment was in fact fairly calm. But before I had even reached the door, the Oxford's port wing went over, the nose dropped and, if I had not hastily scrambled back and straightened her up, we would have been out of control and we would have spun in.

Unless someone else was flying the aircraft, it was difficult to get out of an Oxford. And in weather conditions reported in the Thames Estuary that day, I am sure no one could have got out – especially from a low altitude – unless there was someone else holding the aircraft reasonably straight and level.

Then the look-out from the *Berkeley*, Bruce Waugh, says he thought the pilot was trying to ditch. A pilotless Oxford in those weather conditions would have spun in.

Oxford V3540 appeared all the time under control, breaking up on entering the water. Amy's two bags came to the surface. The pilot was also thrown out, but he would certainly be dead by the time Fletcher reached him.

Who he was, whether he was a man she met at the Hawkstone Park Hotel near Ternhill on that diverted trip, or a secret agent, or a friend is immaterial. It is unlikely we will ever know. People in those times of the blitz simply disappeared without trace.

They must have seen the convoy. They must have made their arrangements. He would stay at the controls, keeping the Oxford level so that Amy could jump with a reasonable hope of survival. Both would know that he would have no chance at all.

Another theory? Yes, of course, but one that fits the facts and evidence more closely than any other. Romantic as she always was, it could be the happy ending to her search, if after her treatment from lesser men, if after all her bad luck in all her love affairs, Amy Johnson finally found a man who possessed something of her own selflessness and courage.

5

The Other
Side of Luck

Charles Kingsford Smith was born 9 February 1897 under the sign of Aquarius, the same as Lucky Lindbergh with more than a silver spoon in his mouth.

He was the seventh child of a bank manager. And as everyone knows, the seventh child has all the luck in the world. That good luck manifested itself early. The family had a half share in a small boat, the *Idalia*. Sailing from Cairns, it was lost with all hands – but neither Charles nor any of his family were aboard. When he was twelve years old, he and three friends set off on a raft from Bondi Beach in suburban Sydney. Swept out to sea by the treacherous undertow, they were eventually rescued. One of the boys was drowned, and Charles was unconscious and almost dead. But a nurse was on the beach nearby and after thirty minutes artificial respiration managed to bring him round. A reckless motorcyclist, he crashed into a store but he was unhurt. Joining the army at the beginning of the first World War, he survived the bloody slaughter at Gallipoli where so many of his fellow Australians were killed. Transferring to the RFC, he crashed on his first solo but survived. At the Front, he accidentally spun a Sopwith Pup, but managed to get out of it. Caught napping by German fighters after a scrap, he should have been shot down himself but, riddled with bullet holes, he managed to escape home with the loss of three toes.

Decorated with the Military Cross, after the war he threw himself into the beginnings of civil aviation.

He had the perfect personality. Small, curly-haired and handsome, he was adventurous, fearless and would take on anything – racing, stunt-flying, joy-riding. Everybody liked him and he had immense charm. Extrovert and never studious, his quick temper was relieved by an almost permanent boyish grin. He cracked up two aircraft, then started airline flying and broke two more. In 1922 he won his first record – 312 miles in two and a half hours. With a friend called Anderson, he was going to fly the Pacific, but their sponsor was drowned before the attempt could be made – which, in the state of reliability of both engines and airframe at the time, was lucky for them. Forced down in the desert by a sand-storm while flying for West Australian Airways, he went on surviving.

In 1924 his flying days almost came to an end. Not liking being employees, he and Anderson bought a service station and began to make money, though he became increasingly unhappy at being earth-bound.

And then in 1927 he had his biggest stroke of luck so far – he met Charles Ulm. Some partnerships appear to be made in heaven – Gilbert and Sullivan, Rogers and Hammerstein, Trippe and Lindbergh, Ulm and Kingsford Smith. No match for him as a pilot, Ulm was a far better businessman. Their combined talents were unbeatable.

Kingsford Smith sold the service station and they bought two aircraft in which they flew the first Australian fare-paying passengers. They flew round the continent, attracting immense publicity and government backing. Now the time was ripe for that delayed Pacific flight.

Together with Anderson, they went to America. There they bought from the Australian explorer, Sir George Wilkins, a Fokker trimotor that had crashed in Alaska called *Detroiter* which they renamed *Southern Cross*. Both renaming an aircraft and flying a crashed one are considered unlucky in pilot lore – and initially it looked as though the whole enterprise would come to nothing, as the government and other backers withdrew and Anderson returned to Australia by ship.

But a wealthy ship's captain called Hancock came to the rescue. With a ship's navigator and radio operator, Kingsford Smith and Ulm set off for Honolulu.

The weather was poor. There were several occasions when the starboard wing (designed to act as a lifeboat in the water) looked as though it would have to be used. But after a number of mirages had disappointed them, Honolulu was seen ahead after twenty-seven and a half hours. Heavy rain lashed down from thunderclouds on Kingsford Smith in the open cockpit on the second leg to Suva. The third leg to Brisbane was just as bad. But there fame, fortune and three hundred thousand people awaited them.

Everyone wanted to back Ulm and Kingsford Smith. Together they bought five more Fokkers, all named after a sky inhabitant with *Southern* in front of it. Together they formed Australian National Airways and with pilots of the calibre of Jim Mollison, Shortridge and Scotty Allan, they were soon outflying Qantas, their only real rival.

In extending their boundaries, there were further brushes with death. Trying for a round-the-world flight in *Southern Cross* in 1929, they were forced down in the West Australian desert as they had lost their wireless aerial ninety miles out of Sydney and could not receive the warnings of the storms that overwhelmed them. A massive search was organized to find them, in which aircraft from the carrier *Albatross* took part as well as their old friend Anderson, flying a Widgeon bought with £1000 that Kingsford Smith had given him.

Then Anderson too disappeared. Now for the first time gossip touched the smiling hero. He was reported to have married a tempestuous Irish girl who had soon deserted him. Worse, the disappearance of *Southern Cross* was reported by two witnesses as a stunt to whip up publicity. Worse still, though Kingsford Smith, Ulm and their two crew were found safe and their aircraft eventually flown out unharmed, the Widgeon crashed and Anderson and his companion were both killed.

Kingsford Smith was hissed on his return to Sydney. His reputation wavered – but only momentarily. The inquiry cleared his name, though criticizing some aspects of the flight. A month later he flew *Southern Cross* on a record flight to England in twelve days and eighteen hours. In June 1930, he flew the same aircraft on the first east–west flight across the Atlantic to Newfoundland and then on to New York 'the hard way' against the wind, receiving a huge ticker-tape welcome before flying on to California, thus completing his own circle of the world. Three months later, he broke the England–Australia record in the single-engined *Southern Cross Junior* with a flight of nine days and twenty-two and a quarter hours.

By then his Australian National Airways had built up an accident-free record over their expanding network, with eighty percent of their flights departing and arriving on schedule. They were the obvious choice of the Australian government to become the equivalent of Britain's Imperial Airways.

Luck he certainly had had – and his luck was clearly continuing. Like seamen, pilots are superstitious and many carry lucky charms. There was a strange toy monkey with a long tail on board the airship R101. Alcock and Brown carried 'Twinkletoes' and 'Lucky Jim' – a hero of the breakfast cornflakes commercial 'High in the air jumps Lucky Jim, Force is the food that raises him'. In the RAF 'gremlins' were wicked little gnomes that threw spanners into the works, and many rites were performed to exorcise them before take off. Before leaving Ireland for his flight across the Atlantic, he had filled the tanks of *Southern Cross* with 1298 gallons instead of its capacity of 1300, remarking to spectators, 'No use borrowing trouble with unlucky thirteen.' Perhaps he was remembering Hinchliffe who had taken off to fly the same route on 13 March two years previously against his better judgment – and had promptly disappeared.

No such fate awaited the lucky seventh child. Just further success and a golden limitless future. On Saturday 21 March 1931 Kingsford Smith won the Segrave Memorial Medal, presented by the Royal Aeronautical Society, the Royal Aero Club, the Automobile Club and the Institute of Aeronautical and Mechanical Engineers for accomplishing 'the most outstanding demonstration of the possibilities of land, air and water transport and stimulating others to uphold British prestige by demonstrating how courage, initiative, skill and the spirit of adventure can assist mechanical development.'

And then on that same Saturday, Australian National Airways' most senior pilot Travis Shortridge arrived at Sydney airport with the lucky gold sovereign in his pocket that he never flew without to take *Southern Cloud*, sister ship of *Southern Cross*, to Melbourne.

Shortridge had been called out to replace a pilot who was on an emergency flight with serum for a sick child – and a sudden change of crew has always been considered unlucky by airmen. His engineer-co-pilot was 23-year-old Charles Dunnell.

There were six passengers. Charles Hood was a producer at the Capitol Theatre, Sydney and he was flying down to spend the weekend with his pretty actress wife who was starring at Melbourne's Theatre Royal. Bill O'Reilly was an accountant. Julian Margules was an expert in the new idea of talking pictures and was returning to his wife and baby after a business trip. Hubert Farrall was in the cream-distribution business, and that day looked very prosperous in his big tweed overcoat, sporting a gold Rolex watch on his wrist. May Glasgow was housekeeper for a medical specialist and his wife and was returning to their home after a holiday. The second woman passenger was Claire Stokes, an artist, wearing a string of chunky lapis-lazuli beads and on her first flight.

The weather report that Shortridge read in the *Sydney Morning Herald* showed 'a marked tropical influence. The present indications favour unsettled weather generally with more rain and thunderstorms, further heavy falls inland with some flooding in low-lying areas. North-east to north-west winds will change during the weekend to squally westerly and easterly winds'. The news page reported a railway disaster on the Paris–Bordeaux express: five people killed in the dining car.

Neither item was particularly cheerful. Two passengers had cancelled at the last minute so as to extend their holiday. That meant two empty seats and £19 less in fares.

At exactly the same time as his fellow ANA pilot Allan left Melbourne for Sydney in the third of ANA's five Fokkers, *Southern Moon*, Shortridge took off from Sydney in *Southern Cloud* for Melbourne.

Watching her go, Claire Stokes' boyfriend had a strange premonition that he would never see her again.

Two hours and twenty minutes afterwards the telephone rang in the ANA office at Sydney. Ulm picked up the receiver. It was the meteorologist with the latest Melbourne weather.

'The city is experiencing the most severe storm for thirty years. On the cliffs, it is impossible for people to stand, the wind is so strong. Large trees have been uprooted, roofs lifted off and telephone poles blown down. Trains are delayed, and beaches strewn with wrecked fishing boats.'

Ulm looked across at Kingsford Smith.

'And the weather en route?'

'Following heavy falls of snow near the headwaters of the Tumut river at Kiandra and Kosciusko, there are fears of floods. Creeks and tributaries are swollen by rain, and by afternoon, the river will be running a banker. Instead of northerlies, winds will be southerlies at around sixty miles an hour.'

Instead of a wind behind him, Shortridge would have a hurricane dead on his nose.

Southern Cloud had no wireless. Radio beacons en route were non-existent. There was nothing anyone could do. But standing there powerless to help, Kingsford Smith must have been reminded of the time three years before when Sydney had tried to warn him in *Southern Cross* of the storm approaching Western Australia that had forced him down.

Initially, there was no panic. Shortridge had gone through worse storms unharmed. Melbourne must expect him to be late arriving.

The minutes ticked by. Allan arrived at Sydney after a very fast flight. He had been in cloud all the way and had been blown well east – but *Southern Moon* had got through perfectly safely.

After lunch, passengers' relatives and friends began to telephone. At first simply routine inquiries, and they were told *Southern Cloud* would be late. By mid afternoon when there was still nothing, worry began to grip. Mrs Shortridge, the pilot's wife, sitting in a cinema, suddenly felt the need to get home. A certainty that something had happened to his son seized the co-pilot's father, Mr Dunnell.

Evening came down, and there was still no sign. *Southern Cloud* had fuel for only nine hours. Kingsford Smith and Ulm comforted themselves with the thought that Shortridge must have landed to refuel at Bowser or some other landing ground. And he had not been able to get through to tell them because the telephone lines were down.

That was the theory which they clung to for most of Sunday. Allan flew back to Melbourne, looking out for any sign of the missing aircraft. The weather was still bad, and he saw nothing. But the Bowser theory was exploded – word came that Shortridge had not landed there.

The Monday edition of the *Sydney Morning Herald* of 23 March contained little but disasters. The *Royal Scot* had jumped the rails at Leighton Buzzard, killing six. The liner *Montclare* had hit the rocks. Two South African Air Force planes had collided, and one pilot had been killed. The remains of an Italian aircraft had been found on a mountain.

And there in heavy type in the centre page:

MISSING
PLANE SOUTHERN CLOUD
EIGHT PERSONS ABOARD

And below: 'Delayed by severe storms on Saturday afternoon, the Australian National Airways airliner *Southern Cloud* carrying six passengers and two pilots, is lost in the north-eastern district of Victoria.' This was the wilderness of the Australian Alps, with mountains like Mount Kosciusko going up to 7314 feet.

The company hoped that Shortridge had been forced down by blinding rain and had been unable to let them know. The search began with all available aircraft.

Now the sightings came in. They were reported as having landed twenty miles east of Violet Town. They had been seen over Goulburn, at Bowser, at Wangaratta. Numerous people had seen them over the heavily timbered Yea district, about forty miles north of Melbourne.

Kingsford Smith searched the Tumut area of southern New South Wales, Allan combed the Yea district. Hope was still high. But the possibility was now published that 'in flying low to retain sight of the ground, *Southern Cloud* might have struck a hill or trees on high ground'.

At Boho, it was seen twice. A ground party set off to search the Whittlesea and Kinglake district north of Melbourne.

ANA suspended all its services 'until the missing plane is found'.

Tuesday came, and there was still no sign of *Southern Cloud*, though dozens of people reported seeing it all over the place. 'The last definite report,' said the *Herald*, 'was from the Flowerdale area where a farmer heard it passing over about 4.30 pm on Saturday,'

Flares and mysterious flashes of light on the hills near Pleasant Creek were believed to have come from *Southern Cloud*.

Kingsford Smith was almost never out of the air. Coming in to refuel at

Holbrook, *Southern Cross* nosed over in wet bog and crashed. He was soon airborne again with Allan and Mollison, searching Beveridge, Glenburn, Bradford, Tallerock.

Every available aircraft was despatched from Melbourne. Relatives and friends spent their time at the airport vainly begging each pilot for news on their return.

A bonfire was reported on the hills north of Kinglake. *Southern Star* took off with six observers. A lady at the Glenburn Hotel, twenty miles south-west of Yea, said she saw *Southern Cloud* flying very low, two or three miles north-west of the hotel. Trappers north of Flowerdale saw the machine at 4.30 pm on Saturday.

The Civil Aviation Department now said that so many reports had been received that it was established 'beyond question that the machine reached a spot within forty miles of Melbourne before it was obliged to descend'. This area narrowed to a densely forested strip twenty miles long and fifteen miles wide.

Fourteen aeroplanes and a thousand men set off to search. The weather now cleared, and as *Southern Moon* flew low over the high plateau round Mount Disappointment, the observers could see for miles.

There was still nothing.

The pilots took enormous risks. Allan in *Southern Star* dipped to within inches of the ground. 'The country became more rugged,' one observer reported. 'The timber increased in height and density, and the hills gave way to

The region of south-eastern Australia which was searched for
Southern Cloud.

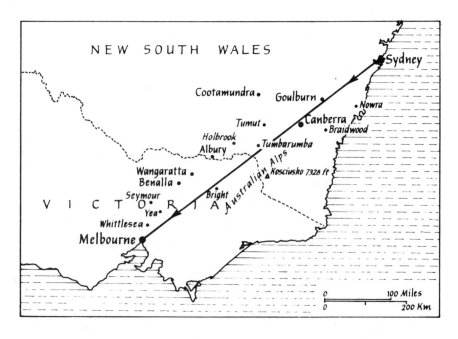

mountains, the summits of which rose far into the low dense clouds. The sides of the hills and mountains were extremely steep, and the valleys between them were hundreds of feet deep. Skirting the sides of the hills at times so closely that it was possible for the observers to distinguish individual fronds of bracken, the *Southern Star* virtually passed through tunnels, the sides of which were the mountains and the tips of which were the clouds.'

A reward was offered for *Southern Cloud*'s discovery, and now it seemed to have been seen everywhere. No systematic search was made. Ground parties and pilots seemed to dash off to investigate every report. Lights were seen at Tintaldra, flashes from the Toolong range towards Mount Kosciusko. Two gold fossickers, prospecting between Braidwood and Nowra, had heard a terrific crash on Saturday. 'I have never seen anything like it before, as far as rough country is concerned,' said a pilot who had searched the Tumut and Cootamundra area.

Intense interest followed an announcement by Robert Byatt of Meragh Station, twenty-five miles from Tumbarumba. Two men called Dyer and Cross, who were three miles from each other on Saturday afternoon, had told him that they had heard a plane pass overhead at that time. Later there was a crash, followed by an explosion. Particular reliance was placed on the report, since Byatt had no idea that an aeroplane was missing, and only heard the news when he returned to the Station on Wednesday 25 March.

A search party set off towards the Yellow Bay Mountains.

Reports continued to come in from Albury, Tumbarumba, Snowy River where a man called Mead saw a plane at three o'clock. Others saw what appeared to be Very lights in the direction of Black Mountain, a range on the Snowy River.

A party of searchers led by Constables Cook and Williams scoured the deep gorges round Nerriga, but were forced back by the impenetrable curtain of gum trees and scrub.

Had the *Southern Cloud* overflown Melbourne and come down in the sea? A large piece of metal was reported in the clear water of the channels, but all the investigating boat could locate were two sharks.

Had it fallen in the Yea district where there had been most reports? Or at Tumut? Or Albury? Or Mount Kosciusko? Or Mount Disappointment? Had it turned east? Or west? Or had it turned back to Sydney?

Kingsford Smith believed that either Shortridge, not realizing the strong headwinds, had imagined he had reached the flat country round Melbourne and had descended into a thickly wooded mountain, or he had deliberately overflown the city to let down over the sea and the storm had driven him into the water.

Gradually the searchers lost heart, though reports of sightings were still coming in, each one of which was being conscientiously investigated.

Australian National Airways, which had been so well on the way to establishing itself, sagged under the combined onslaught of the cost and the publicity.

Kingsford Smith and Ulm were at their wits end. 'It is regrettable,' said Ulm, 'that sensation-mongers will create and disseminate harmful rumours.'

The weeks went by. The rumours increased, but there was still no sign.

Kingsford Smith's luck had been turned upside-down.

Round his neck now was a dead albatross called *Southern Cloud*.

The first thing to go were his flying records.

C.W.A. Scott was fast approaching Australia from England in a Gipsy Moth on which was emblazoned his wife's name in Chinese characters. On the day that the inquiry into the missing *Southern Cloud* opened in Sydney, Scott arrived at Darwin, having smashed Kingsford Smith's record in *Southern Cross Junior* by more than eighteen hours.

Then aeroplanes Kingsford Smith had owned and flown in began to crash. The inquiry had no sooner started than *Southern Cross Junior* disintegrated over Sydney, killing pilot and passenger.

The members adjourned to go and look at this new disaster. On their return, the inquiry did not go well. The only report Shortridge was stated to have received officially before departure was 'windy and rainy weather'.

Responsibility for taking off rested with the pilot, Ulm said.

'Do you,' he was asked, 'as a pilot think there is any likelihood of a pilot in very adverse conditions taking off in spite of them, on the belief that if he didn't, he might be regarded as a "cold-footer"?'

'No,' Ulm replied.

But as the schedule stood it was as though two pilots were pitted against each other at exactly the same time. If one left from Melbourne, the other would leave Sydney, and vice versa.

Ulm further maintained that the aircraft could fly on two engines, even cover a hundred miles on one. The shortest time on the route had been two hours forty minutes, the longest seven hours. Training, particularly on blind flying, was supposed to be good. The pilots flew about 1200 hours a year, which ANA considered not excessive.

It *was* excessive. Civil airline pilots today average half that on aircraft equipped with excellent autopilots and every sort of radio and navigational aid.

What is extraordinary was that Kingsford Smith and Ulm stated publicly that the two most important things they had learned from their Pacific flight were the necessity for radio (they had one) and the need 'not to trust to luck but on a reliable and experienced navigator'. They should have insisted on both (even though radio signals over the mountains might be weak) on a route that made even Jim Mollison, later to shatter so many world records, tremble and dub it as one of the most dangerous in the world.

All the inquiry came up with was to reiterate those two most important things that Kingsford Smith had already learned three years before.

The Sydney–Melbourne service was stopped till there was proper radio coverage. But the reports on the *Southern Cloud* did not. A message on a piece of card saying, '*we are hopelessly lost. Compass done. Shorty*' was alleged to have been picked up on a New South Wales beach. New 'evidence' was continually turning up in the newspapers. Search parties set off periodically. 'Clues' and bits of pieces supposedly from the aircraft were regularly sent in to the Civil Aviation Department.

The ghost of *Southern Cloud* began to push Kingsford Smith nearer and nearer towards his own grave. ANA lost £10,000 on the search. All route flying within Australia ceased.

Bravely he fought against the turned tide of his luck. He set off in a single-engined Avro Avian *Southern Cross Minor* from Australia to try to beat Mollison's Australia–England record, but was forced by bad weather to land on a beach north of Singapore. Though he managed to get off again, he suffered further difficulties such as engine failure, becoming unconscious, making a navigation error, then became sick and was detained by soldiers in Turkey, finally arriving in England days behind Mollison's record. Still sick, he was told by doctors that a return flight was out of the question, and he sold the plane and returned by boat.

Trying to get in first on the Australia–England airmail service and revive ANA, Scotty Allan set off with Christmas Mail in *Southern Sun*. But he crashed trying to take off from a waterlogged aerodrome in Malaya. Kingsford Smith left Mascot to pick up the mail, and struck a telegraph post at Darwin.

The mail finally did get through, and seventy thousand letters were brought back to Australia. At long last Kingsford Smith was knighted. But those two flashes of the old luck were not enough. ANA were finally liquidated. Qantas, not ANA, became Imperial Airways' partner on the Australian run.

The golden boys – Ulm the businessman and Kingsford Smith the flying genius – split up. The new knight was reduced to joy-riding and barn-storming. Ulm scraped up all his resources to buy *Southern Moon* and rechristened it *Faith in Australia* to try for the Atlantic east-west record. Loaded too heavily with petrol, it crashed taking off from Portmarnock Sands. Repaired again, it knocked eleven hours off the England–Australia record, only just won again by Kingsford Smith on *Miss Southern Cross*. Next year he first established a New Zealand airmail service.

Then Kingsford Smith and P.G. Taylor flew the first west-east crossing of the Pacific in a single-engined Lockheed Altair *Lady Southern Cross*.

Those records – and they had established similar ones in the past all over the world – had boosted ANA in the past, collecting publicity, money and fame. Not any more. Those days had passed. Now airlines were just on the point of operating the world routes. The Australia–England airmail – Ulm's original idea – would be flown by Qantas Empire Airways in partnership with Imperial Airways.

Doggedly, Ulm and Kingsford Smith still fought to get back on top. It was as though they were both trying to fight back the malign influence of *Southern Cloud*. For the last four years, Kingsford Smith had been regulary reviled by the Australians who had placed him on the highest pinnacle. Cowardice, not illness, had stopped him trying for the England–Australia record in *Southern Cross Minor* – that had been the first calumny. Now he was accused of being unpatriotic for choosing an American aircraft for his east-west Pacific flight, though there was no suitable British alternative.

There were regular rumours and headlines in the newspapers, but *Southern Cloud* had still not been found. The bad luck continued. And now it was not just Kingsford Smith who was affected. The bad luck appeared to have spread to the aeroplanes he flew and the people he met and had dealings with.

The record is quite incredible even for those times. The boy who died with him on the raft, his crashes in the war and afterwards, his friend Anderson killed looking for him when *Southern Cross* went down at Coffee Royal. Captain Holden who found them not long afterwards was killed in a crash.

Then came *Southern Cloud* and the missing eight. Then *Southern Cross Junior*, killing pilot and passenger over Mascot in Sydney. Then the extraordinary story of *Southern Cross Minor*, the aircraft he was alleged to have been too cowardly to fly back to Australia, which had been bought by Bill Lancaster.

RFC pilot, steeple-chaser, boxer and bronco-rider, Lancaster had left the RAF in 1926 and tried to fly to Australia with the dark-haired Mrs Chubbie Miller, wife of an Australian journalist. Their Avian had crash-landed on an island off Sumatra. There they remained for months, being overflown by Hinkler, the first to fly in a light plane to Australia.

Not surprisingly, they had fallen in love. On their return, Lancaster took his wife and two children to America on a flying job, and Mrs Miller took a bungalow in Miami. Mrs Lancaster and the children soon returned to England, and Mrs Miller obtained her divorce. But Lancaster could not obtain his freedom. The depression had begun, and Lancaster roamed America looking for jobs. He met Amelia Earhart and frequently flew with her.

Meanwhile back at Miami, Chubbie Miller had fallen in love with a so-called airman/author named Haden Clarke. They both wrote to Lancaster to tell him they were going to be married. '*Congratulations to you both, but not till I return to be best man,*' was his telegraphed answer.

Lancaster returned to Miami on 13 April 1932 and the three of them drank and talked over the situation on the bungalow's verandah. Eventually they retired – Chubbie to her room, locked on Clarke's orders, the two men to single beds on the verandah.

In the middle of the night, the discarded lover was pounding on Chubbie Miller's door with the news that Haden had shot himself with Lancaster's revolver. A suicide note was by Haden's side, but a handwriting expert declared it had been written by Lancaster, which Lancaster freely admitted.

Not surprisingly, Lancaster was put in gaol, charged with first-degree murder. Not surprisingly, his defence attorney at first refused to take the case, on the grounds that 'the man is as guilty as hell'. But he changed his mind. So did the judge who took the case. Haden was shown to be a drug-addict, and to have two wives already. Lancaster had kept a diary of his mental torment, and the judge said, 'It has been my privilege to see into the depths of a man's soul through his private diary . . . I have never met a more honourable man than Captain Lancaster.'

Acquitted, he decided to attack the England–Capetown record, once held by Kingsford Smith's old employee and record rival Jim Mollison, but smashed by his wife Amy Johnson, and now standing at three days seventeen hours.

For the purpose, Lancaster bought *Southern Cross Minor*. On 11 April 1933, seen off by his parents and Chubbie Miller, he left Lympne with a religious poem, a silver horseshoe, Chubbie's photograph, a bar of chocolate, two gallons of water and a packet of sandwiches made by his mother.

He reached Oran, hours after Amy's time. He was later still at Adrar. Hope of the record had practically gone at Reggane, but at 6.30 that evening he still took off for Gao.

Since then, nobody had set eyes on him. Like *Southern Cloud*, two years before, *Southern Cross Minor* simply disappeared. It was considered that Lancaster had either deliberately crashed and killed himself, or Bedouins had captured him and made him their slave.

Next to go was *Miss Southern Cross* in which Kingsford Smith had beaten the England–Australia record. She broke up over Sydney, killing the passenger and severely injuring the pilot.

Then in December 1934, Ulm set off from Oakland to beat the US–Australia record in an Airspeed Envoy, which I and hundreds of other RAF trainee bomber pilots flew during the war under the name of Oxford.

Ulm in *Stella Australis* was heard round Hawaii, asking for a fix. But their reply must have been drowned in static. Ulm's last message was '*Fuel exhausted. Coming down in the sea. Please come and get us.*' Nothing more was heard or seen. The Envoy's ditching qualities were poor and it simply disappeared. Navigator and pilot were more separated than Earhart and Noonan.

Now the old veteran *Southern Cross* nearly followed suit. Flying the King George V Jubilee airmail to New Zealand, one of the engines packed up. Shortly afterwards, a second engine began losing a dangerous amount of oil. P.G. Taylor undoubtedly saved the aircraft by repeatedly going out on to the wing and transferring oil from the dead engine to the sick one on the other side – but they had to jettison everything, including the mail.

Short of money, Kingsford Smith tried to sell 'the old bus' as he called it to the Australian government at a bargain price. He had been unable to sell *Lady Southern Cross*, still lying in California after his Pacific record. There was no alternative but to ship her back to England preparatory to flying her to Australia, having a crack at the new record on the way.

In England, he again fell ill. The Australian government had agreed to buy *Southern Cross*, but now he could not get the money out of them. A further effort to sell *Lady Southern Cross* failed.

In spite of the fact that his passage by ship was already booked and he was too unfit to fly, he was literally frog-marched into the air by a whole army of ghosts and previous circumstances. Short of cash, short of the honour due to him for his vast contribution to world-wide aviation, he *had* to fly *Lady Southern Cross* in record time to Australia to win back his old prestige, throw off the curse of the still missing *Southern Cloud* and restore his fallen fortunes.

Kingsford Smith and an engineer called Pethyridge, who was devoted to him – as his employees always were – set off from Lympne at the end of October 1938. Within a couple of days they were back again, beaten by storms over the Aegean.

Saying that this was to be his last major flight, he and Pethyridge set off again at 6.27 on 6 November, shortly after Broadbent and Melrose, each in Percival Gulls and both out to beat the record.

Refuelling at Athens, a Dutch flyer called Van Dyck warned him that there had been gales for over a week in the Akyab region. But on he went over Asia Minor, the Persian Gulf, India – he had no option.

He made his usual perfect landing at Allahabad. Tanked up again, he took off for Singapore just after 9 pm local time.

High above the Bay of Bengal at 3 the next morning, his fellow Australian Melrose saw *Lady Southern Cross* flash past his Percival Gull in the darkness, two hundred feet above him. Kingsford Smith was going flat out at around 220 mph, twice Melrose's speed. 'It gave me an uncanny feeling,' he said, 'over the desolate ocean to see spurts of flame coming from the twin exhausts of the *Lady Southern Cross*.'

That was the last time anyone saw Kingsford Smith. *Lady Southern Cross* disappeared to join *Southern Cloud* and *Southern Cross Minor* in oblivion.

On 9 November, the *Sydney Morning Herald* announced in big headlines:

KINGSFORD SMITH

FEARS FOR HIS SAFETY

LOST BETWEEN AKYD AND VICTORIA POINT

Melrose had waited in vain at Singapore for Kingsford Smith to turn up before abandoning his own flight to Australia to look for him. Two RAF flying boats joined him, then a Vildebeeste bomber squadron. Two days later, they were still searching.

'It is not likely,' said Kingsford Smith's older brother, 'that the boy flew past Singapore. My theory is that he took the jungle route in Burma to avoid the monsoon storms over the ocean. It is also possible, if he has been forced down, that he may have landed on one of the emergency airport fields which we established along the Malay peninsula, on the coastline of the Bay of Bengal and which are without radio communication.'

A different view was held by Littlejohn, who had just completed a honeymoon trip from England to Australia, and described the route from Rangoon to Alor Star as 'absolutely rotten with dense jungles, mangrove swamps and occasional bits of beach'. The weight of water falling from a rainstorm a fortnight before had forced him down to just above the sea. Intensity of storms could be gauged by their colour, he said. Grey ones were not too bad, black storms were serious, but brown ones were lethal.

'My opinion,' he went on, 'is that if an aircraft, travelling at 200 miles an hour, were driven through one of these storms, either the engine would be torn out completely or the fuselage and the wings would collapse. In which case, the aircraft would simply crash into the ground, a complete wreck.'

'My way,' he said, 'was to go as slow as possible.' In this , he was quite right. Normal procedure flying big airliners fifteen years later was to do just that. Even so, Constellations would emerge from the Bay of Bengal or from storms over India with eight hundred rivets shaken loose.

Broadbent, who had preceded Melrose and had arrived at Darwin after breaking Kingsford Smith's record by seven hours and forty-seven minutes, said he had flown through 'terrific gales known as Sumatras', the worst weather he had ever been through. 'No more record attempts for me in single-engined planes,' he said. 'I feel quite numb. You could stick a pin in me and I don't think I would feel it.' Once he had felt so drowsy that he had thrown a thermos of squash over his head to keep himself awake.

The search for Kingsford Smith continued. And now Melrose went missing.

Not surprisingly, the Australian government considered the discontinuance of record-breaking flights in view of the danger.

Now it was *Southern Cloud* all over again. Rumours were rife. Everyone seemed to have seen *Lady Southern Cross*. Everyone had their theories. An aeroplane had been reported flying low near a tin mine at Takuata at 7.30 am on 8 November. A squadron of bombers proceeded to the area.

Kingsford Smith's old comrades P.G. Taylor and Scotty Allan had joined the search. Ground parties were hacking through impenetrable forest. Notes in English, Malay and Chinese were dropped to jungle dwellings. Flares were reported off the coast of Siam by a passing steamer. Footprints and boat marks were discovered. A Siamese train driver had seen something. So had a villager at Kjupum.

Melrose turned up, having landed on a beach and bent his propeller. He had tugged it through the forest and by train to Penang to be repaired, then returned by plane to fit it and take off again to search.

Gradually and inevitably, as with *Southern Cloud*, the search slackened. Finally after landing at Darwin on 28 November, Melrose, who had given up his chances of the record to look for Kingsford Smith, now said there was no hope of him being alive. Six days later, he who had already crashed once, crashed again. Fog forced him lower. He saw a green patch. He said to himself, 'Here goes!' and pushed the nose down. He couldn't remember anything after that. Unconscious for days, he suffered severe concussion.

As though for buried treasure, people still went on searching for *Lady Southern Cross*. Reports were still coming in of 'sightings' and 'finds' on 1 May 1939 when a whole wheel and oleo leg was washed up at Ate Island just off the Burma coast. On the tyre was stamped: '*Goodrich Silvertown Aeroplane Type B 2621 Serial 5841*'.

Covered with shellfish, it had certainly come from Kingsford Smith's Altair, but the seaweed that might have given an indication of where in the ocean it had been lying had been cleaned off. All the marine biologists could say was that it had been resting undisturbed on a muddy bottom at a depth not exceeding fifteen fathoms.

Again the RAF took up the search. A party of divers and experts in ocean currents left Australia to join them.

So it was to go on over the years, as men still searched for *Southern Cloud*, *Southern Cross Minor* and *Lady Southern Cross*.

In 1953 wreckage was found on Mount Disappointment, but it was that of an RAAF trainer. Two years later, more wreckage – identified eventually as that of a US dive bomber. Three years later, a New Zealander radiesthetist (someone who can receive messages from the ether and contact missing people) offered his help to a radio broadcaster who was running a programme to locate lost relatives and friends. The radio broadcaster sent him a newspaper photograph of a woman and a map of Victoria and asked him to pick out where she had last been seen.

Near Tumut, came back the answer, and woman was a passenger in a plane crash. The photograph in fact was that of May Glasgow, one of the two women on *Southern Cloud*.

But still nothing was found.

On 26 October 1958 a man called Harold Dymond was just setting off to look for the missing aircraft at exactly the same time as a young carpenter employed by a firm on a contract to build a dam on the Tooma River was standing in the forest round Tumut in New South Wales, trying to get a picture of Mount Kosciusko, just to the south. It was Tom Sonter's day off, and he was a keen photographer.

Suddenly the ground gave under him. Looking around, he saw rusty wires twisting through the branches of blue gum trees. Through green shadows, he glimpsed the tops of five seats.

A brass petrol tank top sent out a dull gleam from the wet scrub. He picked it up and read, '*Avro Type X. Capacity fuel 78 gallons*'. Another gleam from the grass – the name plate from a key chain which read, '*Clyde E. Hood, Capital Theatre, Sydney*'.

He went back with his finds to the contractor's camp. Someone said a plane was supposed to have come down round here many years ago. There was some talk that it was carrying gold, and Sonter set off to see a solicitor in Sydney. He had never heard of *Southern Cloud*, which had crashed before he was born.

But rumours of his find had preceded him. He was intercepted by the police, and a search party including a Ministry of Civil Aviation expert was organized.

They had to wade through icy rapids, clamber over broken rocks, dodge small landslides and falling stones. Cutting their way through a mass of undergrowth, standing at first light on the freezing mountainside above the roar of the creek a thousand feet below, they saw the plane in the tangle of tree trunks.

After twenty-seven years, *Southern Cloud* had been resurrected. Not much of it certainly – there had been a fire at some time, consuming most of the wings and fuselage. Shortridge had been turning to the right and had got on to a north-east course back to Sydney. By a stroke of bad luck he had struck the top of a ridge above Deep Creek. Kingsford Smith, Allan and the others had flown many times round here – again it had been bad luck that they had seen nothing. The regular services linking Melbourne and Sydney, now ten a day, continually passed not far away, but they too had seen nothing. It had certainly been *Southern Cloud* that the shepherds Dyer and Cross had heard, and again it had been bad luck that their vital clue had not led anywhere. Nor had that of the inhabitants of Tintaldra, only fourteen miles away, who had heard the plane.

Farrall's gold Rolex was found – stopped at 1.15, along with several sovereigns, a tiny scent bottle, a man's studs in their box, a woman's bead necklace, a metal clip for a suspender, a razor, broken binoculars, a thermos flask still containing liquid, and the bones of the pilot's foot still in their shoe.

At least the discovery brought peace of mind to the relatives. Mrs Hayter, Shortridge's daughter, and Mrs Stokes, mother of Claire, had both been haunted with the idea that their loved ones had been wandering around suffering from loss of memory. There had been so many false reports, so many searches. Each one had brought the tragedy back again for them.

'Every time a wreckage was found,' Mrs Hayter said, 'the story of *Southern Cloud* would be revived. Naturally we are not happy, but there is a sort of relief. At least we know they did not suffer.'

Mrs Ewart, sister of Mrs Farrall, said they went through hell. She and one of her brothers had gone to Tintaldra soon after the crash to check on reports. 'We had 500 million people telling us of 500 million places where the plane had been either seen or heard. There was no central organization to plan the search, so we did it ourselves. We even went to Canberra to see the Prime Minister. The people who did help – the storekeepers, the postmasters – were marvellous.'

What *had* happened?

Shortridge, after flying blind into worsening weather, could have turned back. At least he would know the weather behind him and he would be getting away from the mountains. Given the same circumstances, a present-day pilot would never have gone in the first place. If he *was* caught by the weather, then back he would go. It is an interesting insight into the character of Kingsford Smith and most of those early pioneers – one is reminded of Exupéry and the relentless push-on manager of *Night Flight* – that to them going back was equivalent to retreating.

It was Jim Mollison, the most famous of Kingsford Smith's employees, husband of Amy Johnson, who exactly described what had happened to *Southern Cloud* twenty years before it was found. In his book *Death Cometh Soon or Late*, he visualized Shortridge flying blind round Mount Kosciusko:

He decides to turn and run back as best he can to the flat country round Canberra. Suddenly out of the clouds, an isolated peak looms up before the pilot: he tries a steep turn but it is not steep enough: his wing catches the hillside . . .

Just over three years after *Southern Cloud* was found, suddenly out of the mists of time *Southern Cross Minor* reappeared. In February 1962 a French army motorized unit, patrolling south of the desert atomic station of Reggane, came across the skeleton of an aeroplane in the Tanezrooft region of the Sahara, known to the Bedouins as the land of thirst.

Close behind was the mummified skeleton of Bill Lancaster, only forty miles from the road which was his route. Why he was not found within days remains a mystery.

There was also his diary of twenty-nine years before telling exactly what had happened. 'No one is to blame. The engine missed. I landed upside down in the pitch dark, and there you are.'

He treated his injuries. He prayed. He debated whether he should stay with the wreck or strike out across the desert. But he had promised Chubbie he would stay and stay he did, as day by day the vultures circled closer.

He worked out his philosophy. 'To be loved,' he concluded, 'is to exist.' He wrote letters to his parents and to Chubbie. His mouth was so dry that he could not swallow the last of the chicken sandwiches.

The last diary entry reads, 'So the beginning of the eighth day has dawned. It is still cool. I have no water. No wind. I am waiting patiently. Come soon, please.'

Now only what had happened to Kingsford Smith remained a mystery, and that tantalizing piece of undercarriage was still the only clue.

Theories were aired. Searches continued. The distinguished Australian airman Sir Lawrence Wackett considered the Pratt and Whitney engine in the Altair had failed. It was an earlier model, and later the speed ratio of the supercharger impeller had been reduced from 12:1 to 8:1. Kingsford Smith must have been going flat out for the record to be flying at twice the speed of Melrose. That would precondition a failure in a marginal engine.

The writer Pedr Davis, continuing this theory, believed that Kingsford Smith had glided down dead-stick over Aye Island with his undercarriage down, intending to land on the beach. The lowered oleo leg had struck palm trees and had been cut off, falling into shallow water near the island, eventually being washed in by the tide. Supporting this, there was evidence that palm tree tops had been fractured on a path to the sea, in which case the rest of *Lady Southern Cross* should be in shallow water close to the island. Yet subsequent searches have so far failed to locate it.

Another theory is that Kingsford Smith fell asleep at the controls. He was unfit before he left England, and had had very little sleep en route. Then flying at 15,000 feet without oxygen, both he and Pethyridge might have passed out.

Or he might have flown at high speed into a brown cloud and broken up. But in that case, how could the floating wheel sink to a depth not exceeding 15 fathoms on a muddy bottom?

It is unlikely he ditched. Had he done so, he would have pancaked down with his undercarriage retracted in the wings. It would have then been impossible for the wheel and oleo leg to fracture and come floating to the surface.

Every few years another party goes off to solve the mystery. As the years go by, people will argue and theorize and search for the truth about what happened to *Lady Southern Cross* and why Kingsford Smith's luck so suddenly changed.

But long ago, Mollison, who had accurately visualized what had happened to *Southern Cloud*, also divined the hinge of Kingsford Smith's luck. He symbolized it by buying a Mew Gull, painting it red and black for death, and calling it *Southern Cloud*. In so doing he may also have been divining his own death. Within a year of the finding of *Southern Cloud*, Mollison had died, a shaky old man at the age of fifty-four, swallowed up by his playboy life of wine and women.

Perhaps Mollison was the aircraft's last victim. For victims there certainly had been. No other pioneer aviator left, through no fault of his, such a trail of death behind him as Kingsford Smith: the boy on the raft; his friend Anderson and Hitchcock, killed looking for him at Coffee Royal; Holden, the man who found him; Shortridge and Dunnell and their six passengers in *Southern Cloud*; pilot and passenger in *Southern Cross Junior*; the pilot injured and the passenger killed in *Miss Southern Cross*; Bill Lancaster in *Southern Cross Minor*; Ulm and his crew in *Stella Australis*; finally he and Pethyridge in *Lady Southern Cross*.

Yet no one did more for world aviation. Today the Melbourne–Sydney route is as safe as houses, and Qantas, the rival that ousted Australian National Airways, is top of the world's airline safety league. And at long last his 'old bus' *Southern Cross*, the last survivor, is preserved as Kingsford Smith's monument at the airport of his native Brisbane, while the international airport in Sydney, from where Shortridge set off that fateful day, has been renamed Kingsford Smith airport.

Was *Southern Cloud* the hinge of Kingsford Smith's luck? Or was that long string of subsequent disasters simply a coincidence?

6

The Arms of Coincidence

On the night of 29 April 1952 a Pan American Boeing Stratocruiser arrived at Rio de Janeiro from Santiago with forty-one passengers on board. The crew of nine 'slipped' – that is, a new crew took over – and after refuelling, Captain Grossack at 02.09 taxied to the end of the runway.

There Rio tower called him. 'Cleared off airways direct from Rio to Port of Spain, Trinidad. To cruise from Rio to Barreiros check point at 12,500 feet. After that to Santorem at 14,500 feet. And from Santorem to Port of Spain at 18,500 feet.'

The first officer reported back the clearance. Two minutes later, the Stratocruiser was airborne into a clear starlit night. The weather forecast was excellent. It would take them ten and a half hours to cover the 2350-mile distance.

The Stratocruiser in many ways was the ultimate in aviation luxury. Designed to compete with the splendid Cunard Queens and the other massive ocean liners that were still carrying the great majority of passengers over the world's routes, it had two decks – the lower deck being a comfortable lounge bar. In the main deck were sixteen bunks, each having their own dividing curtains and equipped with linen sheets, pillows and blankets. Pan American and BOAC vied with each other to provide super-service, and passengers were provided with free cocktails, free gifts, dinners of caviare, turtle soup, cold Inverness salmon, spring chicken with Wiltshire bacon, strawberries and double cream, cheese, frivolities, fresh fruit. Free champagne, free cigars, free wines and spirits, extra-special stewardess service.

However, in some ways, the Stratocruiser was a retrograde step to the Constellation that it was designed to supersede. It had an ugly pig-like snout, huge 3500 h.p. 'corn-cob' engines (so called because the thirty-six cylinders were arranged in four banks which resulted in inadequate air-cooling), was bad on three engines and impossible on two, and was altogether too heavy to cruise in the stratosphere that its name pretended.

That night Captain Grossack maintained a north-easterly course over Brazil. The passengers who had bunks were asleep. In the downstairs lounge others were talking and drinking. At 06.15 the captain reported abeam Barreiros flying in the clear at 14,500 feet, estimating abeam Carolina at 07.45.

95

That was the last message ever heard.

On 1 May USAAF aircraft found the aircraft 887 nautical miles north-north-west of Rio. The wreckage was scattered over a mile of jungle in the notorious Matto Grosso area.

The problem was getting to it through dense undergrowth. A plan of campaign was organized on 4 May to cut a trail, using natives led by Pan American employees familiar with the jungle, from a base camp at Lago Grande. The American government agreed to send a bomber with a helicopter on board to land at a landing strip to be cut out of the jungle at Araguacem.

The Brazilian Boundary Commission and the Indian Protective Agencies told the Americans that though friendly Carajai and Tapirape Indians would be found initially beyond Lago Grande, the wreckage was located many miles deeper in a jungle area occupied by the hostile Ciapos Indians. They advised that any party going should be well armed, should not attempt any contact with the Ciapos, should remain in a group and not become isolated, should fire upon Indians they encountered, should equip themselves with suitable clothing to protect against jungle briars, thorns and vines, and that protection should be provided against wild boars, black leopards, jaguars and snakes of the boa-constrictor and viper species. The area surrounding the wreckage had never been explored, and as far as was known there was no habitation west of the Araguaia River.

The base camp at Lago Grande was established, and the clearing of the Araguacem strip began. Then parachutes were seen on the tops of the trees. An unauthorized Brazilian rescue party had landed and had made their way to the wreckage before the official government expedition.

Eventually Pan American employees reached the wreckage. There were no survivors. The aircraft had plummetted down out of control and the fuselage was scattered over a mile area. There was no sign of Number 2 engine – the port inner.

There were then problems with the unauthorized rescue party as to who had the right to proceed with the investigation. Meanwhile Pan American engineers tried to find out what had happened.

Sabotage was ruled out – there was no sign of an explosion. There was no indication of fatigue failure on the fuselage. There had been no fire. Rudder boost malfunction, flutter and buffeting were all considered and rejected. Bird strike was suggested but no known local bird flew at 14,500 feet.

They then looked at the question of propeller failure. Two years before a propeller failure had occurred in a Stratocruiser, causing the engine to fall free. On 29 March 1951, after unusual vibration, Number 1 engine of a Stratocruiser began drooping in its nacelle, and after heavy shaking twelve and a half inches of propeller fell off. Number 2 engine of the Brazil crash was never found. But the most likely cause was considered to be propeller failure and steps were taken by the manufacturers to strengthen them.

Some time later another Pan American Stratocruiser was flying over the Pacific when intense vibration was encountered. The captain had turned back to the USA, when Number 3 engine and propeller tore loose and fell free, the aircraft becoming almost uncontrollable. Unable to maintain height, the captain ditched thirty-five miles off the Oregon coast. All twenty-three

1a   The York aircraft

1b   The R101 at its launching mast

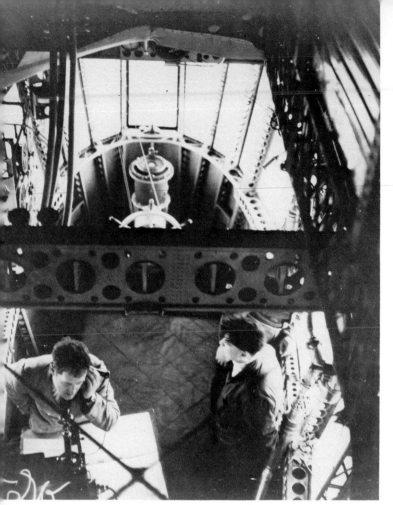

2a　Inside the R101's control cabin

2b　Squadron Leader E.L. Johnson, Sir Sefton Brancker, Lord Thomson and Lt Col. V. Richmond, all of whom died in the R101's crash

3a   Flight Lieutenant H.C. Irwin

3b   Eileen Garrett

3c   Air Chief Marshal Dowding

3d   Muriel, Lady Dowding

4a   An aerial view of the R101's wreck

4b   The French poacher Alfred Rabouille giving evidence at the R101 inquiry.
The model shows the airship's angle at the time of the crash.

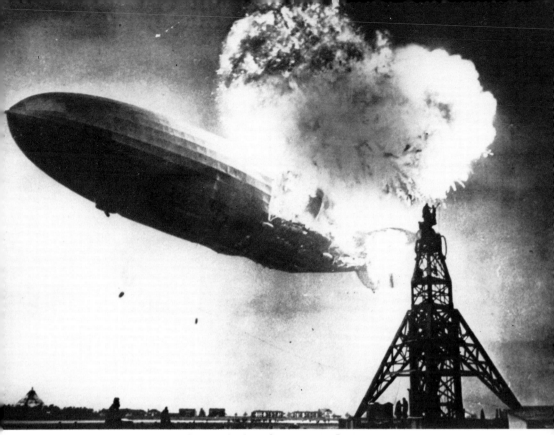

5a   The *Hindenburg* bursts into flames

5b   A B314 flying boat of the type in which Churchill flew across the Atlantic.

6a   Amelia Earhart

6b   Amy Johnson

6c   Amelia Earhart, photographed from the navigator's seat, at the controls of the plane in which she and Fred Noonan disappeared.

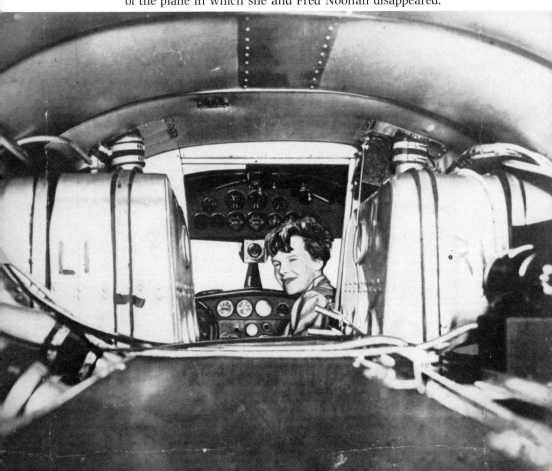

7a   Amy Johnson and her husband Jim Mollison

7b   Kingsford Smith (right) and Ulm (left) being congratulated on their arrival
at Croydon from Australia

8a   Charles Kingsford Smith in *Lady Southern Cross* over Oakland, California
at the end of his flight across the Pacific

8b   The last photograph of Kingsford Smith, taken as he took off from
England for Australia

9a   A Boeing Stratocruiser

9b   Captain H. Gulbransen

9c   Dag Hämmarskjold

10a   General Tshombe

10b   General Sikorski

10c   A Liberator aircraft

11a   A De Havilland Comet I

11b   Captain Harry Foote

12a   Captain Norman Macmillan

12b   The Fairey Postal Monoplane

12c   Flight Lieutenant H.H. Jenkins and Squadron Leader A.G. Jones-Williams

occupants were evacuated safely, but there was one serious injury and four people died later. The aircraft sank in water a mile deep.

There were a few further incidents involving propellers and constant speed and feathering mechanisms on Stratocruisers. On 6 December 1953, at 10,000 feet half-way across the Pacific between Honolulu and Wake Island, sudden vibration was felt by the first officer in the control column of Pan American Stratocruiser 90947.

The flight engineer and the navigator promptly checked the engines and the spinners for rough running – but everything appeared normal.

The steward was on his way up to report unusual vibration in the galley when he saw a flash of fire on the starboard wing.

Up front there was a sudden building up of vibration culminating in an explosive noise and a violent jolt. Next moment, number 4 engine (starboard outer) and its propeller fell away.

This exposed the large flat plate area of the firewall to the airflow, and with her skin shaking all over in violent buffeting, the aircraft dived out of control in a right-hand descending turn.

Full left aileron and rudder trimmer tabs were rolled on. Flap was tried, but made no difference.

With the combined efforts of both pilots applying full left rudder and aileron, control was regained over the aircraft. But the right wing still would not come up, and as a result the aircraft was uncontrollable and went on turning.

The first officer broadcast 'Mayday' (the radio emergency call). The passengers put on their life-jackets and were briefed for ditching.

The captain now made an unusual and courageous decision. There, in the middle of the Pacific, he jettisoned 2500 gallons of petrol from the starboard wing and so managed to lift it.

The heading of the aircraft was at last controllable within 20° at a height of 2300 feet at a speed of 145 knots, and course was set for Johnston Island, well south of track, where a safe landing was made.

After that accident Pan American instituted more extensive propeller checks and began to replace hollow propellers with solid.

Three years later another 377, N90943, half-way across the Pacific between Honolulu and San Francisco, and close to the US weather ship *Pontchartrain*, was cleared to 21,000. On reducing power at that altitude, suddenly number 1 engine (port outer) oversped. It was found impossible either to bring the revolutions down or to feather the engine, and the captain decided to cut off the oil supply and 'freeze' it. But the propeller still continued to windmill and the aircraft was losing height at the rate of 1000 feet a minute.

Maximum continuous power was put on both inboard engines and partial power on number 4. Control was regained, but there was no chance either of going on to San Francisco or turning back. The passengers were prepared for ditching by the weather ship.

It was found that the windmilling propeller could be kept under control at an airspeed below 140 knots, and for five hours N90943 circled. During this time number 4 engine failed and had to be feathered. At dawn *Pontchartrain* laid a foam path on the water on a heading of 315°. The sea was calm, but there was a heavy swell.

With full flap and undercarriage retracted, the captain approached the water levelled off at 90 knots. Touchdown was light, but a tremendous impact followed as the aircraft decelerated and began to sink.

The fuselage broke aft of the main cabin door and crew and passengers scrambled to safety into twenty-man rafts, all being picked up by the *Portchartrain*.

The cause of the accident was considered to be a malfunction of the constant speed and feathering mechanism of the propeller control system on number 1 engine.

A year later another Pan American Stratocruiser, the next number in line, N90944, was over the same weather station between San Francisco and Hawaii, and reported flying 10,000 feet at 17.05.

It was the last message the aircraft sent.

Five days later, bodies and wreckage were spotted from the air ninety miles north of the aircraft's track. All nineteen wore life-jackets – clearly the passengers had been prepared for ditching. There were no survivors.

N90944 had solid propeller blades. Certainly a ditching had been attempted – which had failed.

The two big questions were why no message had been sent, and why was the ditching made away from the weather ship and ninety miles north of track?

The fact that there was no message might mean that some part of the radio equipment had been damaged, possibly by a propeller spinning off into the fuselage, possibly followed by the whole engine, as had happened in the earlier Stratocruiser that had diverted to Johnston Island. Then the aircraft had been uncontrollable until the captain jettisoned fuel. If he had not, the aircraft would have careered many miles off course. But jettisoning fuel was a most unconventional thing to think of when you are in a steep diving turn half way across the Pacific.

The inquiry came up with no definite cause. But, historically, one or other or a combination of the causes of the accidents to N90947 and N90943 would seem the most probable factor in the accident to N90944.

BOAC did not have the problems with their propellers that Pan American had. On Christmas Day 1957, at the same time as the investigation into N90944 was still going on, Captain Val Croft arrived at London airport to fly BOAC Stratocruiser G-AKGM across the North Atlantic.

Balloons, paper hats and novelties decorated the passenger cabin. The traditional Christmas dinner of roast turkey, plum pudding and mince pies was served with champagne. Coffee and liqueurs were being served half-way across the Atlantic, when suddenly there was a shudder right down the passenger cabin.

The steward went up front to report vibration aft, but on the flight deck, Captain Croft was well aware of it. His lighted torch was directed out of the side windows on to the darkly turning propeller blades.

He was looking for the shadows of desynchronization. There were none. All the blades marched perfectly in step. Then his eyes shifted to the four oil pressure gauges. On number 4 engine – starboard outer – the needle had fallen very slightly behind the others.

Fourteen thousand feet below was the December Atlantic. Even from this height could be seen the glint of icebergs in the sliver of a waning moon. In such freezing conditions, human beings would last ten minutes at the most. And they were 500 miles from land.

'Oil temperature's rising on number 4,' the engineer officer reported.

The vibration had increased. Now the whole instrument panel was shaking. Captain Croft took out the automatic pilot and brought the throttle lever back on number 4. The juddering continued.

'Looks like we'll have to feather number 4,' Croft was just saying to the engineer officer when suddenly there was a shriek from the starboard wing.

'Number 4 over-speeding, sir!' the engineer officer shouted.

'Bring back the revs!'

'Trying to, sir!'

It was impossible to hear on the flight deck. The rpm needle went off the revolution counter.

'Prop'll come off in a moment, sir!'

'Feathering!' Croft banged the red feathering button forward. 'Feathering number 4!'

But the propeller refused to feather. Instead, up went the revolutions. The nose was swinging violently. Both pilots were struggling to keep her straight.

The second engineer was sent to pump oil into number 4 from the emergency tank.

The first officer monitored it from his side window. 'Oil's just pouring out of the engine! Not doing any good!'

The Stratocruiser had begun to lose height at a thousand feet a minute.

'Report what happened on the R/T,' Croft told him. 'Send Mayday!'

The first officer picked up his microphone. 'This is Speedbird Golf Mike . . . five hundred miles off Newfoundland . . . Number 4 engine over-speeding . . . cannot feather . . . losing height. Mayday . . . Mayday . . . Mayday!'

Then he switched off. There was nothing in his earphones now but endless silence.

'Mayday . . . Mayday . . . *Mayday*! Speedbird Golf Mike reporting number 4 over-speeding and losing height.'

Still nothing.

'Mayday . . . Mayday . . . Mayday,' the first officer went on calling. 'Speedbird Golf Mike . . . propeller over-speeding – '

'Golf Mike . . .'

Very far away, half drowned by atmospheric crackling, came another human voice.

The whine of the propeller was higher than ever. The whole flight deck was shaking and shuddering.

'Golf Mike, what aircraft are you?'

'Stratocruiser.'

'Prop over-speeding?'

'Yes!'

'Golf Mike . . . this is Captain Gulbransen, Pan American . . . I had the same . . .'

Captain Gulbransen had been flying a Pan American Stratocruiser when suddenly intense vibration was encountered, followed by intense buffeting.

Engine oil pressure started dropping. Then a propeller had run away, producing an intense banshee wail. Gu bransen knew the propeller would come off at any moment and ricochet into the fuselage.

The passengers were prepared for ditching. One of them was Danny Kaye who, according to newspaper reports, kept morale up with wisecracks. One of the stewardesses was resting in a cubbyhole behind the downstairs bar. Wakened by the whine of the propeller and hearing the chief steward's voice calling on passengers to leave the lounge and take up ditching positions, she got up and dashed to the door – only to find it locked.

The chief steward, counting heads, found she was missing, guessed where she was and rushed down to the lounge again. Pushing open the door, he told her there was not a moment to spare, rushed out himself and up the stairs – letting the door slam and again imprisoning her. But finding her still missing, he returned to the lounge to check and let her out.

The Stratocruiser limped slowly on.

By the time Captain Gulbransen brought her to safety, he knew all about dealing with a runaway propeller through the best possible school – personal experience.

That knowledge he now passed on to Captain Croft, struggling with Golf Mike far below him over the icy Atlantic.

'Close your gills . . . that'll reduce the buffeting.'

'What about flap?'

'No use.'

'I'm still losing height.'

'Put the inboards to rated power. Throttle back number 1. You'll be able to hold height around three thousand at 140 knots.'

The minutes went by. The Pan American pilot still stayed with the British pilot, still talking.

Shades of those earlier Stratocruisers – the one lost in the Brazilian jungle, the one that sank off the Oregon coast in water a mile deep, the one that just made it to Johnston Island after jettisoning fuel from the right wing, the one that ditched by the weather ship, the one that disappeared ninety miles off track – must have passed over all those passengers strapped to their seats in their life-jackets, waiting to ditch in the icy Atlantic.

Any of those fates could have been theirs – except that their captain now knew what might happen and was more than one jump ahead of possible events.

'Where's your nearest airport, Golf Mike?'

'Sydney.'

'You're on course for it now?'

'Yes.'

'I'll keep in touch . . .'

Captain Gulbransen's presence was the best and most unexpected Christmas present any on board Golf Mike could have had.

Fifty Stratocruisers had been built, flying for eight years many millions of safe miles all over the world. Of the hundreds of pilots who had flown them, those who had had this propeller trouble and were still operating could be counted on the fingers of one hand. Those few pilots could have been anywhere in the world that night. What chance was there that one of them should have been in earshot of Captain Croft?

It was a most extraordinary coincidence.

What coincidence is has baffled scientists for centuries. Does it exist? Is there a pattern or a meaning in it?

A reader of a novel will accept certain coincidences, but will reject others so fiercely that he will stop reading the book. Why? Is belief the thread which binds him to the story?

In Thomas Hardy's *Tess of the D'Urbervilles*, the blood of the false lover that Tess has slain forms a red heart shape on the ceiling. Taken out of context, it becomes totally unbelievable. Blood flowing out of a body simply does not form such a pattern. But within Hardy's emotion-charged story, the reader can swallow it without a tremor.

A writer of novels is conscious of the 'cut-off' action of coincidence on readers and avoids it. Yet coincidences occur in fact that would be rejected in fiction. A novelist would not be able to put into the plot the idea of the water ballast bag bursting above the R101 aft engine car in which Binks and Bell were struggling against the fire, putting it out for them. Nor would he be able to include the Atlantic meeting of Captains Croft and Gulbransen.

The scientist Kammerer and the psychologist Jung kept log books of coincidences stretching over many years. Most of them were trivial – serial runs of tickets, people of the same name meeting within a short space of time. Space and time are certainly connected to coincidences – it is their unity that makes for the coincidence. Two accidents, one after the other, at one place on one night is a coincidence. A dozen accidents occurring at the same time in different parts of the world are not.

Kammerer thought that coincidences came in series – hence gamblers' lucky or unlucky days. Certainly we know that accidents come in series – the superstition of one accident being followed by two more has a basis of fact. Accident-proneness has been shown to be a kind of intermittent wave.

Kammerer believed that alongside casuality – that is, events that are 'caused' – 'is an a-caused principle which tends towards unity in the universe. Caused and a-caused act together to form a world-mosaic or cosmic kaleidoscope, which in spite of constant shuffling and rearrangements, also takes care of bringing like and like together'. He coined the word to describe this as 'seriality'; Jung used 'synchronicity' – 'a coincidence in time of two or more casually unrelated events which have the same or similar meaning'.

Jung put the phenomena down to his theory of mankind – the collective unconscious and the archetypes operating therein. Strong emotion or affinity took over and facilitated the occurrence of a coincidence – that was his view. Einstein related coincidence to two-dimensional time. Eddington believed in a five-dimension universe with three spatial and two temporal dimensions. Dobbs thought the probable world moves through a second-time dimension – events follow probabilities, which would account for precognition but will not do for coincidence.

Koestler points out that all is one and one is all is at the basis of many religions, and this idea of 'unity' is endorsed by Jung, who has stated, 'all precognition and the paranormal are based on the idea that random events are minor mysteries which point to one central mystery'.

'The propeller may come off,' said Captain Gulbransen. 'If it does, it will shoot ahead of you. Take care to avoid it!'

Stratocruiser Golf Mike was now holding height, but the wild shrieking of the propeller over-speeding was still there. Warned of all possibilities, Captain Croft watched and waited.

Eventually the magnesium propeller housing caught fire and burned with a brilliant white light. Then suddenly the big four-bladed propeller itself came off and raced ahead, still madly turning like a giant catherine wheel, before disappearing into the Atlantic night behind Golf Mike as the Stratocruiser caught it up.

Captain Croft reported what had happened to Captain Gulbransen.

'How are things now?'

'Better.'

'Still able to maintain height?'

'Yes. And we're in contact with Sydney. Thanks for your help.'

'That's all right, Golf Mike. And . . . happy Christmas!'

Captain Croft continued on three, making a safe and skilful landing at Sydney, Nova Scotia on Christmas Day.

No other Stratocruiser propeller trouble was reported until fifteen months later. And then there was another extraordinary coincidence.

Stratocruiser G-ANTY took off from Accra in Ghana and began climbing. The report stated,

during the ascent, on number 4 engine the oil pressure dropped and the oil temperature rose 10° above the others. At 8000 feet, following an increase of speed from 170–180 knots, the rpm on number 4 engine rose rapidly. Propeller over-speeding drill was initiated, but the propeller refused to feather. Reaching 7000 rpm the propeller became uncoupled and fell away. Fire broke out in number 4 engine and continued to burn, but without spreading. The aircraft returned to Accra where a safe landing was effected and the engine fire quickly extinguished.

Another dangerous situation had developed which might easily have been fatal.

Except that the captain was Val Croft, who had left BOAC and joined Ghana Airways.

The chances of Captain Gulbransen and Captain Croft meeting together within radio telephony range that Christmas Day over the Atlantic must have been hundreds of thousands to one against. The chance of two number 4 propellers falling off two Stratocruisers under the command of the same captain, with no intervening Stratocruiser propeller trouble occurring, must be almost as remote.

But what about the chance of the same captain figuring in the same two coincidences? What are the astronomical odds against that?

Can that strange meeting between those two captains be explained by Jung's Collective Unconscious? Or by Einstein's Two-dimensional Time? Or by the theories of fate and destiny? Or by some unknown telepathic communication – some strange sense in man that remains as yet subterranean and unrecognized?

7

The
Sixth Sense

Touching, Seeing, Hearing, Tasting, Smelling – those are the five senses, and the evidence appears to be that man is fast losing these powers, not gaining another one. In comparison to animals our performance is poor. We do not seem to inherit instincts as animals do. We have to be taught skills like swimming. We do not come into the world ready equipped with much of the know-how we need.

We learn to see, but we also learn *not* to see. So many stimuli bombard us all the time that we would be overwhelmed if we took them all in. So we are selective, reserving out concentration for our own interests. You have to go for a country walk with a naturalist to realize what you are missing. Then there is a censorship system probably located in the reticular formation that does not allow us to see things that are against our upbringing and education. Recent psychological experiments have shown a preconscious processing within us. Stimuli below perception threshold are *physically* registered which mentally remain unseen. Recent experiments have been done on people who have had operations involving hole in the head surgery. Electrodes placed on the brain have shown registration of stimuli of which the individual's consciousness remains totally unaware. There are capacities inside ourselves, sixth and other senses, that have been allowed to atrophy or have been deprived by conformity or other pressures or have never been properly developed. As a result, in comparison to our enormous perceptual potential, it is as though we go through life like kittens, with our eyes half closed, resisting attempts to open them. Sixth and other senses, even those warning of danger, are thereby distrusted.

But one instinct, self-preservation, we do still appear to possess. Psychological experiments have shown that if you put a baby to crawl on the table, when he comes to the very edge, he will hesitate and stop. As adults, when we come to a cliff edge, most normal people have an overwhelming urge to move away from it.

In the same way, perhaps inherited from flying animal ancestors millions of years ago, when you learn to fly an aeroplane, it comes naturally *not* to get the nose too high. There is again a strange feeling that goes right through you if you do. Somehow you know you will not be able to become airborne. You know instinctively it is wrong. You actually feel the drag on the wings. And you push the nose down.

It is possible that a pupil on his first flying trip might conceivably get the nose too high and fail to correct. It is quite impossible for an experienced airline captain to do so unless there is a very good reason.

Such a man was Captain Foote. In 1952 I was flying piston-engined Constellations on the Atlantic, while the crack new Comets, the first jet airliners in the world, captured all the limelight. Foote was a member of the BOAC Comet Flight, regarded as an elitist group.

Not that Harry Foote was so inclined. I was to get to know him well. A quiet, modest, soft-spoken man, very methodical and deliberate, he had been born in Antigua in 1916, the son of an engineer. His forebears were Irish, and a Foote was one of the first three men to settle on the uninhabited island in 1632. Of medium height, this Harry Foote had brown hair and hazel eyes. He had come over to Britain to fight in the Second World War, had joined the RAF, completed a tour on Lancasters and been awarded the DFC and Bar. Afterwards he joined BOAC and on 26 October 1952 was a senior captain with 5868 flying hours behind him, mostly in command, of which 245 had been on Comets.

At 17.56 GMT on that day he lined up the blue and white BOAC Comet I G-ALYZ on runway 16 (that is, pointing 160° Magnetic) at Ciampino Airport, Rome and called for the before take-off check.

It was pitch dark and raining. The wipers clanked across the windows to reveal muzzy blobs of runway lights, but there was no horizon. Trim was set to neutral, and the flaps lowered 15°. The four Ghost jet engines were opened up to full power, and rpm checked at 10,250 on all engines. Fuel flows, engine temperatures and pressures were reported correct.

Captain Foote released the brakes. Slowly into the damp darkness the heavily-loaded Yoke Zebra moved.

The Comet was the new British wonder plane. Built on a private enterprise venture by De Havilland, sleek, stream-lined, the engines sunk in the wings, it was also one of the most beautiful. Orders had been received from airlines all over the world, and it was confidentally predicted to be a money-spinner. On 2 May 1952 the first ever scheduled jet service had been inaugurated after a wildly enthusiastic send-off from Heathrow. Yoke Zebra was on this same service to Johannesburg, and her next stop was to be Cairo.

The Comet gave a strange banshee whistle when it flew, but inside the cabin all was quiet and vibration-free. Cruising at almost twice the weight of piston-engined aircraft, high above the weather in blue skies, the passengers, at five miles a minute, looked down from five miles high at the miniature world below.

Yoke Zebra began gathering speed down the slippery runway. The needle on the airspeed indicator crept round the dial – 60, 70, 75 knots.

80 knots . . . the speed laid down in the BOAC training manual to lift the nose-wheel. Just as he had been instructed, Foote eased the control column back.

The nosewheel came off the ground. The speed built up to 112 knots, the already correctly calculated 'unstick' speed. Again Foote moved the control column back, this time to lift the aircraft off the runway.

Yoke Zebra inched off the ground. Foote called, 'Undercarriage up!'

At that instant, the port wing dropped violently. The aircraft swung left, then began juddering. Twice Foote tried to correct on the control column, but the juddering continued.

Instead of rising, Yoke Zebra bounced back on the runway.

It was as though there was a voodoo on her. Everything appeared perfectly normal – *but she simply would not fly.*

'Voodoo' aircraft were common, both in the RAF and in civil flying. I never knew a squadron or a line that did not have one. I have flown them and on occasion refused to fly them. My first novel, *The Take Off*, centres on one such aircraft on a Coastal Command squadron, called by everyone J-Jinx. Ten novels later in *Sword of Honour* I told the story of an aircraft at a pilot training school, nicknamed K-Killer.

The RAF blamed bad flying characteristics on goblin-like characters called 'gremlins'. In an earlier age spells and witches caused strange and inexplicable behaviour.

The Comet appeared not to be responding to the controls. Its speed was not building up. And she was rapidly approaching the red boundary lights at the end of the runway. Foote's only thought now was to save the passengers. He slammed back the throttle and tried to stop.

Seconds later, Yoke Zebra was sliding over rough ground. Both undercarriage legs were wrenched off. The wing broke. The aircraft came to an abrupt stop.

There was sudden silence. The reek of kerosene was everywhere. But the crew shepherded all the passengers out safely, and there was no fire.

But there *was* an immediate outcry. Nothing could possibly be wrong with the wonder plane. Hardly before the pieces had been picked up from the ground, let alone examined or an inquiry held, the Ministry of Civil Aviation produced an interim report and BOAC issued a joint statement – all to the effect that they and De Havilland were satisfied that neither engines nor aircraft were to blame for the accident.

That only left Captain Foote. He must have done something wrong. There were certainly tail skid marks on the runway that could only be made if the aircraft was at an $11\frac{1}{2}°$ angle.

On 12 November BOAC held its own court of inquiry. Foote was still somewhat dazed.

Star witness was Captain Majendie, an extremely clever pilot with an engineering degree. Majendie was the Comet fleet superintendent and the oracle on jets.

The story of Harry Foote and the Comet so far was now unfolded at the inquiry.

At the formation of the Comet fleet it became a ritual to have a monthly get-together at a local pub to foster good relations between crews and management. At one of these, Harry Foote, who had just started the training course, said that as the Comet was a totally different aircraft to anything BOAC had used before, it would be a good idea to fly it up and down the routes for training purposes only for at least a year, carrying crews and mail only, as he felt there was a lot to learn about it before passengers were carried.

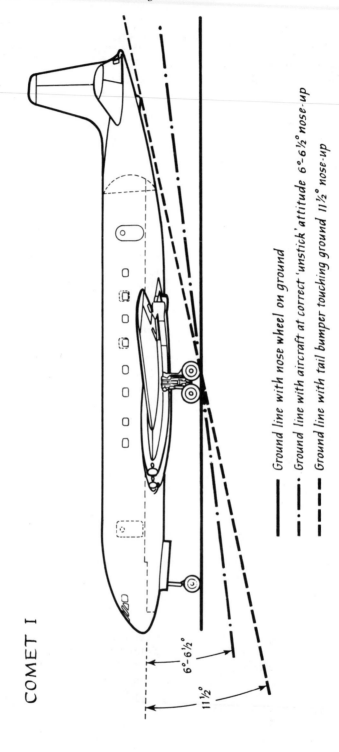

COMET I

6°– 6½°

11½°

—————— Ground line with nose wheel on ground

—·—·—·— Ground line with aircraft at correct 'unstick' attitude 6°– 6½° nose-up

— — — — Ground line with tail bumper touching ground 11½° nose-up

In January 1952 he had started the six-week technical course, then had his first flight, not at the controls but standing between the two pilots on the flight deck. He was then sent on a proving flight to Johannesburg. Since he had not completed his conversion course, he expected to go as an observer, but the captain informed him that he would have to make use of him as he was short of pilots.

The captain was Captain Majendie.

Foote did the take-off ex Rome, Majendie having informed him that there was 'nothing to the take-off of this aeroplane' (Foote wryly noted in his statement that 'in view of subsequent happenings and factors which have since come to light, this seems to have been a doubtfully valid statement'). He also carried out several reasonably good landings.

Majendie expressed his opinion that Foote was 'slow with take-off checks, etc'. Not unnaturally Foote considered this unfair. Far from helping him to familiarize himself with the Comet, the trip had the effect of undermining his confidence. But he took and passed perfectly satisfactorily his Comet (flying) course. He was then sent on the route as captain under supervision. The supervisory captain was Captain Majendie.

There was another captain also under supervision. Foote carried out only one take-off and landing on the trip. On return he was given an adverse report – his first ever.

On his second supervisory trip, when landing at night at Beirut, the flaps struck a fence and were damaged. Considerable difficulty was being experienced at that time with misting windscreens, and the outcome of that inquiry was that the misting and the height of the fence were the cause, not the pilot.

Foote was then sent out under Captain Rodley. The trip went well until, on the return landing at night at Khartoum, unknown to either Rodley or Foote, the runway was flooded and again the flaps were damaged.

Rodley and Foote were suspended. The BOAC accidents investigator arrived. Rodley was subsequently desuspended, but despite Rodley's efforts, Foote was instructed to return as passenger.

The subsequent inquiry held that neither Foote nor Rodley were to blame. Foote underwent a satisfactory six-monthly check, and then completed two trips in total command to Johannesburg without incident but with commendation from a member of BOAC's board.

Then came the Rome crash on his next trip. Foote was found to have got the nose too high, as a result of which too much drag developed for Yoke Zebra to become airborne.

It was a most extraordinary finding. It is quite impossible for an airline captain with 5868 flying hours to get an aeroplane's nose too high and *not* correct . . . unless there was a very good reason.

No reason was found, nor even looked for. Foote was in a dilemma. If he did not accept the finding, he would certainly not be taken back as a pilot by BOAC and would lose his job, and he was trained for no other. No known case of a pilot reversing the decision of a company inquiry existed. The tail skid marks on the runway provided incontrovertible evidence that he *had* got the nose too high. For the life of him, he could not think *how* he could have done, but he could not deny it. But he was sure that he *had* become airborne but the Comet failed to climb away, sank back and then ran along the runway tail down. He had no corroborative backing on this, and the Ministry thought otherwise.

Foote accepted the inquiry's findings. The BOAC Inspector of Accidents then wrote to say that the Ministry of Civil Aviation's Chief Inspector of Accidents was carrying out a parallel inquiry and that, 'As a result I fear you will inevitably be found responsible.' It was a kindly letter, and reassured him that he could keep his licence. In longhand was the PS, '*Don't worry about all this.*'

The Ministry inspector wrote to say he understood 'that as a result of the Corporation's inquiry blame for this accident has been attributed to you and that you have accepted this finding'. The Ministry's findings were likely to be similar, and Foote was asked to sign a government form accepting the total blame. He signed it.

No one had been killed, and the Ministry conducted an inspector's inquiry, not a public one. It was all done quickly and quietly, and in many ways, Foote was relieved.

But retribution followed. On 26 November the Comet fleet manager wrote to Foote, formally telling him that the inquiry had held him responsible, and he was to be admonished. Furthermore, he was to be posted to Yorks, the oldest BOAC aeroplane, used to carry freight and such animals as monkeys, elephants and leopards. Vast publicity in all the newspapers highlighted his punishment and the vindication of the Comet.

For the next few months he soldiered on over the empire routes in Yorks, but he worried about how he could possibly have got the Comet's nose so high. At home, he spent his time doing endless graphs and calculations.

BOAC and De Havilland were worried too. The manufacturers had been doing further tests, and a new take-off technique was introduced. Now the nose-wheel had to be lifted off the ground at the same 80 knots, but afterwards it had to be placed back on the ground again – a most extraordinary manoeuvre. Foote said that when he returned from Rome, he had found amended and undated instructions regarding take-off in his locker.

Gradually Foote was beginning to work out a theory to account for the aircraft's behaviour. When the sympathetic Captain Rodley asked him what his private feelings were now on the accident, he replied that he considered the unstick speeds were too low.

But if that was the case, why was no one else having trouble? There had been 4000 hours testing on the Comet and thousands of hours route flying. Why was it only for Captain Foote that the Comet would not rise?

Confidence in the Comet had been precariously preserved in overseas buyers and on 12 February 1953 Captain Charles Pentland, then Operations Manager of Canadian Pacific Airlines, based in Vancouver, arrived in England to take over *Empress of Hawaii*, purchased by his company, and fly her to Vancouver. I had just come in off a Constellation service that day, and I saw and spoke to Captain Pentland at Paddington station from which we were both going to Bristol, where BOAC's Atlantic line was then based. He was just the same as ever he had been – small moustache, neat, beautifully parted brown hair, a lopsided Clark Gable grin on his khaki-coloured face. I knew him well. So did all the BOAC Atlantic pilots, especially my generation of ex-RAF pilots, for Bill

May and Charles Pentland were the training captains at Dorval for the BOAC Atlantic service. Both men trained us on Liberators and Constellations, passing on all the invaluable tips and techniques learned during countless wartime crossings. But unlike his fellow Canadian May, Pentland was not a born pilot.

About one pilot in ten is born to flow into an aeroplane as though he was its mind and soul. He flies the way a ballet dancer dances, with no visible connection between man and machine. With a 'made' pilot, the join is quite visible. That does not mean to say they are less safe than born pilots – often quite the reverse, because they are more aware of themselves and take meticulous care.

Pentland was a 'book pilot', following everything the Book said and making us, his pupils, do likewise. He used to talk through the side of his mouth, pouring information into us even as we were trying to carry out some complicated flying manoeuvre.

He was very careful and conscientious. Of one thing I was certain. Flyingwise, he would always do what he was told. After all, it was his job to see that other people did what he told them. As a result of his and Bill May's efforts, the BOAC safety record over the Atlantic was the finest in the world.

When BOAC's Atlantic service was returned to England as part of a dollar economy drive, Pentland left the Corporation to take the position of Operations Manager, Canadian Pacific Airlines. He then went to Hatfield where he received instruction on the Comet. He was particularly shown the 'Foote take-off', being warned against getting the aeroplane into that nose-high attitude.

Foote had to go to Hatfield one day and shared a car with Pentland from the factory to the station. Pentland got into the front. Foote sat at the back. Neither spoke a word to each other.

Pentland left London Airport on Sunday, 1 March 1953 with four crew and six technicians, and flew without incident to Beirut, before continuing to India. Since fog obscured Karachi, he diverted to Nawabshab airport, returning to Karachi when the fog cleared.

In the early hours of Tuesday, on a hot dark night at maximum all-up weight, *Empress of Hawaii* prepared to take off for Calcutta.

It sped a reported 3270 yards down the runway (ordinarily Comets are airborne in half that distance), smashed through a barbed-wire boundary fence, had its wheels torn off in a road culvert, slid on its belly over the sandy ground, hit the twenty-foot embankment of a dry rivulet and burst into flames. Four hours later, it was still burning. All on board were killed.

Marks on the runway indicated that the tail skid had scraped the surface. Black skid marks indicated heavy braking.

The accident investigation was carried out by the Pakistan government, assisted by the British inspectors. It ran a bumpy course, with a number of disagreements. At one stage, the British High Commissioner was induced to hold a party to restore harmonious relations.

The report itself was kept confidential. But the finding was an almost carbon copy of the Rome accident report: the crash was caused by the fact that the nose of the aircraft was lifted too high during the take-off run. The partial stalling and excessive drag this produced did not permit normal acceleration, and prevented the aircraft from becoming airborne in time.

Headlines in the papers declared, '*Pilot erred and Comet was wrecked.*' Now another pilot, even more experienced than Foote, a highly competent training captain, was supposed to have made a most elementary mistake, *even after* the 'Foote take-off' had been demonstrated to him. Clearly it was not possible unless there was some other reason, such as that the unstick speeds were too low. Yet the Comets had flown 11,000 hours and no other pilot apparently had reported difficulty.

Foote was convinced that if only he had been listened to, the Karachi accident would never have happened and the eleven souls on board *Empress of Hawaii* would have been saved.

Other people at long last were beginning to share this view. The *Aeroplane* magazine of 13 March 1953 stated that there was

something unusual about the fact that at Rome, an experienced airline captain should have placed his aircraft in an impossible position. It was dark, but he knew the characteristics of the aircraft and the power-operated control. 'Pilot error' could be accepted, in this case, as one chance in a million. But, if the Karachi report suggests that even with the many different characteristics of the case, a similar kind of mistake may have been made, then such a mistake must be presumed to be too easily possible. Whatever the Karachi findings may be, a technical means will obviously be found to prevent a recurrence of the Rome accident and these means should, as far as possible, be explained without reservations.

The *Express* asked for 'a mechanical or an electronic warning system which will instantly tell the pilot if he lifts the nose too high for take-off'.

De Havilland were doing further tests, especially in hot weather. There were rumours that the wing was going to be altered and the take-off technique was to be changed yet again, but De Havilland kept quiet.

Harry Foote was now sure that he had been unjustly blamed. He was also sure there would be more Comet accidents.

Some months later, on 2 May 1953, a BOAC stewardess called Patricia Rawlinson was flying in Comet Yoke Victor on a flight with a full load of passengers from Bangkok to Calcutta.

She was just as worried as Harry Foote. There had been a two-hour mechanical delay leaving, but that was nothing. She had been feeling listless and nervous and suffering from a skin allergy, and had only returned from sick leave the previous week. She had talked to another Comet stewardess before leaving London.

'We sat in her flat having tea,' Irene Milburn reported, 'and I asked her if there was anything worrying her.'

'No,' Pat Rawlinson said, 'at least nothing I can put my finger on. I have a terrible feeling of apprehension which I just can't seem to shake off.'

She had gone off in that frame of mind. Then many small things had gone wrong on the trip. She had still felt unwell, and the steward had been ill. To her fiancé, she had written, 'Things have gone wrong all along the line. There's an absolute hoodoo on this trip.'

The hotel where the crew stayed in Bangkok was run by a Siamese woman educated in England. It was her custom to come with her staff of young Siamese waiters and maidservants to say goodbye to every departing crew and wish them good luck.

The Siamese have an old superstition that if any part of a warrior's armour falls off or is dropped before he goes into battle, he will be killed. Before Pat Rawlinson's crew left that morning, as usual, the Siamese, all smiling, were standing in the grounds at the gates of the hotel to say goodbye.

As Captain Maurice Haddon was about to board the crew car, one of the three gold bar epaulettes of his uniform fell to the ground. None of the Siamese moved to pick it up. They had kept their eyes averted.

Nevertheless the flight so far had been smooth. They had caught up a little lost time. Unknown to Pat, as they approached the Indian coastline, they flashed past Harry Foote in his lumbering old York, twenty thousand feet below.

Twenty minutes after Yoke Victor landed at Dum Dum airport, Calcutta, Foote came in to Operations and had a chat with Haddon, who was making up his flight plan to Delhi.

The weather at Dum Dum was good, but a thunderstorm was expected from the north-west in about an hour's time.

Maurice Haddon, with his five crew and thirty-seven passengers, took off in Yoke Victor at 10.59 GMT, just as Harry Foote and his three crew got into the transport to take them to the Calcutta hotel where they were night-stopping.

At 11.02 Haddon contacted area control: 'Departed from Calcutta 10.59 hours. Estimated time of arrival Palam 13.20 hours. Climbing to 32,000 feet.'

An hour later, when Foote and his crew reached their hotel, they heard a rumour that area control had lost contact with Yoke Victor.

Foote said immediately, 'It will have crashed.'

Next day BOAC issued a statement to the effect that the last message reported the aircraft to be

climbing on track. We do not consider the Comet to be lost since it might have force-landed at some place where there is no means of establishing communications. All planes travelling over the route have been keeping a constant look-out, and Dakotas of the Indian Air Force and a BOAC York freighter took off at first light this morning . . .

Harry Foote in his 'punishment aircraft'. He had been quite right, BOAC quite wrong. Not long after dawn, he sighted wreckage at the village of Jangipara, twenty-five miles west of Calcutta. It was a strange coincidence that it should be Foote who found the crash. The man who had been demoted from the wonder plane now flew round its wreck.

Twenty BOAC and Indian civil aviation officials headed out for Jangipara in cars. Six miles away, they found the road water-logged and had to walk.

A boy gathering mangoes had seen a bright red flash and then a wingless plane plunging. One witness had heard four deafening explosions and the whole sky lit by fire. Scattered around burning mail bags were a child's bonnet, a gay parasol, a woman's skirt, a man's pipe. The first body found was that of Pat Rawlinson. She and the other five crew were killed. So were all the thirty-seven passengers.

What had happened? There had certainly been a storm but a KLM Constellation had gone through it without damage.

At the inquiry at Calcutta, eighteen days later, it was established that the aircraft suffered complete structural failure and was on fire in the air. There was nothing to suggest any faulty material.

The Ministry inspector said there were no signs of engine failure. Prolonged study of the wreckage for anything between nine and twelve months would be necessary at the Royal Aircraft Establishment at Farnborough to establish the sequence of the structural failure, which was due to over-stressing resulting from either severe gusts or from over-controlling, or loss of control by the pilot when flying through the thunderstorm. One assessor considered it possible that the elevator failed through too fierce a 'pull-up' by the pilot.

Why wasn't the structure itself the number 1 suspect immediately? The Indian eye-witness saw the red explosion; if this was so, the aircraft was not inside the storm. Haddon was a highly experienced pilot who would have avoided a dangerous cloud build-up. If the Comet broke up in such weather, what was it doing flying in India where such storms are not infrequent?

The wings of Comet II and Comet III aircraft were modified to improve the lift of the wing. Thirty-four Comet IIs and eleven Comet IIIs had been ordered, but the Americans would not issue a certificate of airworthiness to the aircraft. However, the chairman of Boeing had taken a look at the Comet and decided his company must certainly build a jet airliner – much bigger, with podded engines and greater lift wings.

Foote felt sicker than ever that no one would take any notice of his view that there was something wrong with the Comet. For many months he had been under strain, operating the old freighters in the hot and dusty Indian and African routes, and was feeling quite certain by this time that he had been unjustly blamed for the Rome accident.

He now felt ill on service, and was flown home from Lagos. He was off flying and under his local doctor, who reported him to the BOAC doctor as suffering from 'anxiety'. Another bogy now raised its head. Because of this diagnosis, the insurance company refused to continue to insure his airline transport pilot's licence – which added to his anxiety still further. Since anxiety was the cause of his illness, and if it went on much longer he would lose his licence and his job, Foote resolved that he would attack the anxiety at its roots.

In the light of what he had found out, he was sure that there was something wrong with the Comet – on take-off certainly to do with the unstick speed and the stall.

He went to see Sir Victor Tait, vice-chairman of BOAC, on 11 November and said that two facts about the stalling speed of the Comet I had come to light since his accident. The first was that if the angle of attack of the wing at unstick is increased by $2\frac{1}{2}$–3°, the result was a partially stalled wing, and secondly the aeroplane stalled at higher speed near the ground than in the free air. Both facts were unknown at the time of the Rome accident. The margin of safety then must have been minute.

Tait did not admit the validity of his argument, and therefore Foote, never a union man before, asked for the matter to be brought up at the next central board meeting of the British Airline Pilots' Association.

There is no situation for anyone more painful or more emotive than to be unjustly blamed. The emotion is intensified if nobody believes you. It reaches fever pitch if an aeroplane like the Comet is continuing to fly with characteristics you consider to be dangerous and yet nobody in authority will listen.

Incidents concerning the Comet were continuing. On 26 June 1953 a Comet hit a wall at Dakar. Three weeks later one ended up in a ditch at Bombay. On 25 July a Comet was damaged landing at Calcutta. There were more such incidents in the tiny Comet fleet than in any other BOAC fleet.

It had been his wife Chris's suggestion that he should occupy his time while he was sick by filling in four scrap-books with details of Comet accidents – those that had happened and those that were to come. In his number 1 scrap-book he pasted newspaper cuttings of his own accident, of the *Empress of Hawaii* and Yoke Victor, but he made no comments beside the cuttings.

The torment in Harry Foote's mind can be seen in the scraps of calculations, the notes, the information he collected, perhaps above all in those strange four brown and yellow scrap-books he had bought, one of which was now filled with cuttings of Comet disasters. Ever since his own accident, he had thought of little else.

Further tests on the Comet take-off by the manufacturers now revealed that even the second take-off amendment, whereby the nose-wheel had to be lifted and returned to the ground again, was inadequate. It appeared that the stalling speed with the aeroplane on the ground at take-off increased much more markedly with weight than had been anticipated, and once it had stalled, the Comet would not leave the ground unless the speed were increased very considerably. It was found that instead of 19 per cent margin between power on stalling and take-off safety speed supposed to exist at the time of the Rome accident, the real margin was only 3 per cent (4 knots). This was totally insufficient, bearing in mind the fact that the pilot had no means of judging fuselage attitude on a dark night once the nose-wheel had left the ground.

But why hadn't there been any more Comet take-off accidents after Karachi?

The reason was simple. The 'born' pilots had used their sixth sense and the 'made' pilots now realized that the thing to do was to ignore the traditional take-off procedure and *not* raise the nosewheel at 80 knots, in spite of the wear and tear and the awful rumbling that technique produced. This unauthorized technique had now been made official, and cables sent worldwide to the effect that there was now a *third* take-off technique: the nose-wheel was to be returned firmly on the ground until not more than 5 knots below the unstick speed.

At the same time as the British Airline Pilots' Association requested the Minister of Civil Aviation to reopen the inquiry into Harry Foote's Rome accident, Harry Foote opened his number 2 scrap-book with a newspaper cutting headed, '*35 killed in Comet. Ten children in plane lost at Point Calamity.*' It was accident number 4.

At 9.31 am on Sunday 10 January 1954 Captain Gibson on Comet Yoke Peter took off from Ciampino Airport, Rome – where Foote had crashed fourteen months before – and started climbing into a calm sunny sky, bound for

London. At 9.50 he had sent out a routine position report. And a few minutes later, he had said on the R/T, 'Did you get that – '

Gibson broke off in mid-sentence. Though Rome went on calling him, there was no reply.

A fisherman in his boat said, 'I was fishing when I heard the whine of the plane above the clouds. Then I heard three explosions, one after the other. For a moment all was quiet. Then several miles away I saw a silver thing flash out of the clouds. Smoke came from it. It hit the sea. There was a great cloud of water, but by the time I got there all was still.'

It was very like Jangipara. No detailed structural inspection had yet been completed of the wreckage of Yoke Victor. That crash had been put down to a storm. But the weather over the Mediterranean was perfect.

Alan Gibson and Harry Foote were both former training captains on Hermes and were close friends.

The usual ideas of the cause cropped up again. Fire in the engines? Unlikely. Sabotage? MI5 were brought in to investigate.

Structural failure was considered unlikely, though now the words 'metal fatigue' were being mentioned.

BOAC now grounded all Comets. The frigate *Wrangler* was sent to search for the wreckage with underwater TV and salvage gear. Medical reports on the bodies of the victims established that an explosion had taken place inside the Comet. Nevertheless BOAC issued a statement that after a careful examination, no structural weakness could be found and training flights would be resumed.

Meanwhile the pilots themselves were trying to find out what was the matter with the Comets. There had still been no response from the Ministry to their request to reopen the Rome Comet inquiry.

A meeting of the Comet Sub-committee of BALPA's Air Safety and Technical Committee took place on 24 February 1954. Two representatives of the Comet fleet had been invited, but they had written to say they felt unable to attend. The Committee considered their absence 'a very great loss'.

The Committee were concerned that there had been considerably more incidents in the Comet fleet over its year and a half of operations than in any other fleet during the corresponding period. Some eight BALPA members – all experienced pilots, with hitherto accident free records – had been involved in accidents for which they had been disciplined. Furthermore there was reason to suppose there were further incidents which had not resulted in disciplinary action. There had been half a dozen landing incidents which had resulted in damage and an accident to a Canadian military Comet.

Not surprisingly, the Comet Sub-committee achieved little. Comets were still flying, but no passenger services were being operated. The investigation into the Calcutta crash had revealed there had been an explosion in the rear of the aircraft. No structural weakness had yet been found.

Majendie had been corresponding with the Ministry on the Foote case, insisting that the aeroplane must have been in a tail-down attitude because of the skid marks and the evidence of an eye-witness.

The Ministry now told Foote that the case was particularly difficult because the Comet fleet were 'unable to provide any of the training notes or other information in force at the time of the accident.' As Foote put it in his notes, 'I said they should be made to.'

A meeting between BALPA and BOAC (including captains of the Comet fleet) took place on 6 April. During this meeting BALPA referred to 'the aura of secrecy which surrounded all aspects of Comet operations', and commented how unfortunate it had been that 'the Comet representatives had seen fit to stand aloof from the discussions' at BALPA. Sir Victor Tait said the Association did not appear to have very adequate grounds to challenge the safety of the Comet, at which BALPA began to list aircraft accidents and dates. The meeting advanced nothing.

On 15 April there was a meeting between BALPA and the Ministry. The Ministry asked Foote, who was present, if he had queried the take-off speed figures when he flew under supervision to Singapore with Captain Majendie. Foote said he had used the normal technique in use at that time, and had he been using the wrong technique, he would have expected to have been informed about it. Both the Ministry and BALPA said they had not been able to get a copy of the training notes in force at the time of the accident from BOAC.

BOAC and the British aircraft industry were going through a particularly rough period. The new Britannia turbo-prop airliner crashed into the Severn mud near Bristol. A BOAC Constellation turned on its back coming in to land at Singapore, killing thirty-three.

Comets were again flying on the long-distance routes – and on the first one to Johannesburg was Captain Majendie, who had resigned from BOAC to take up a business appointment at Smith's Instruments Ltd.

The press enthusiastically hailed the news that Comets were flying passengers again. Foote pasted a cutting from the *Evening News* saying, 'The decision to fly them is entirely justified,' into his scrap-book.

On the next page he wrote, '*5th Accident 8 April 1954.*'

BOAC's Yoke Yoke, carrying fourteen passengers and a crew of seven, had taken off *again* from Ciampino Airport and had sent one message, 'Over Naples. Still climbing.' Nothing more was heard.

The circumstances of Yoke Yoke's crash were almost identical with those of Yoke Victor at Calcutta and Yoke Peter off Elba. Shortly after take-off, the Comet exploded.

RAF Shackletons operating from Malta sighted bodies and wreckage, and guided naval ships and fishing boats to the scene where they searched the whole area by means of flares and searchlights.

At Farnborough, tests to destruction were still going on. Out of seventeen Comet Is, six had been written off with the loss of 110 lives. The *Telegraph* said, 'The blame for what now seems to have been a premature decision must rest with the official technical advisers of the Minister. These are the Air Registration Board, the Air Safety Board, his own Ministry and the Ministry of Supply.'

Meanwhile, that same minister had been considering the reopening of Foote's inquiry. There had been some sympathy for the pilot amongst officials, but there had also been an authoritative letter from Majendie, the expert who knew far more about Comets than the officials. Also, there were fifty-four Comets on order, many of them from hard currency countries.

Some of the blame for allowing the Comets to return to service so soon inevitably rested with the Minister and his officials. It would be embarrassing to him if the inquiry was reopened and it was shown that *again* they were wrong.

The Chief Inspector of Accidents wrote to the Airline Pilots' Association informing them that, as a result of his re-examination of the circumstances, he could see no reason to reopen the inquiry.

It was a bitter blow to Harry Foote. With great fortitude, he resolutely went back to flying the Yorks. But his hate of the Comet and those responsible intensified.

Only the scrap-books – he was now on number 4 – kept him sane, according to his wife. There were numerous press cuttings to paste in. The airworthiness certificate had been taken away from the Comet. Pilots volunteered to fly Comets with escape hatches and parachutes day after day on what were called 'death climbs'. But now they were calling for Maxaret brakes throughout the fleet, which would also assist the accelerate/stop during take-off, the fitting of a stall-warning device, the provision of a tail parachute, the fitting of a droop-snoot wing on the Comet II, and the modification of the nose-wheel oleo leg to prevent hammering at the high ground-speeds required for compliance with the recommended take-off techniques. In Parliament Sir Robert Perkins asked John Profumo (Parliamentary Secretary, Ministry of Civil Aviation) if he could reopen investigations into the Rome accident. This was denied but there were loud cheers when Perkins nevertheless proceeded, 'in view of the fact that, since this accident there have been three changes[1] made in the take-off technique of the Comet, will he institute a full public inquiry to enable this innocent man to clear himself?'

Meanwhile, a vast water tank had been built under the brilliant direction of Arnold Hall, the Director of Farnborough. Comet Yoke Uncle had been placed in the tank and enormous pressures were continually exerted on it. And most of the pieces of Yoke Peter had been dredged up from the Mediterranean, brought to Britain by warships and jigsawed together on a frame in a Farnborough hangar. It was the most thorough aircraft investigation ever.

In the water tank, structural failure in the pressure cabin occurred near the rectangular automatic direction finding window. And it was due to metal fatigue. Sir Arnold Hall in court put Yoke Peter's window, retrieved from the Mediterranean, beside the failed window of Yoke Uncle, and showed the same crack in both.

Strengthening throughout and circular windows were put in hand by De Havilland for the Mark II and III Comets. With extraordinary courage, the company continued doggedly on with the Mark IV. Comets resumed carrying passengers, though it was significant that now the wing was one which gave more lift, and special stall warning devices were installed – bells and lights and eventually even stick pushers which pushed the nose down – now became standard on most jet aircraft.

Reports on the inquiry were as usual pasted in Foote's scrap-book. He never allowed himself to comment but now and again he underlined a sentence that had touched him deeply. In the inquiry report, he underlined a question put to the Chief Aeronautical Adviser to the Ministry by the QC who appeared for the

---

1 In fact, there had been two.

relatives of the dead: 'Looking back on it, do you think that this expression "everything humanly possible has been done" is putting it much too lightly?'

The adviser was asked whether he was satisfied that the Calcutta accident was purely due to weather conditions, or whether he had any doubts that it might have been due to some defect in the aircraft. He replied they were not certain about it.

'And yet,' the QC asked, 'you recommended the minister that they should be allowed to fly again?' Again the adviser agreed. The Comet I went on the scrap-heap. But Phoenix-like over the years the new Comets IIs, IIIs and IVs built up an excellent reputation for safety and reliability.

A tiny cutting of 1 September 1955 put in the scrapbook stated,

A four-engined British Overseas Airways Corporation York freighter flying to Bangkok made an emergency landing at Rangoon yesterday after the propellers of two engines had fallen off. It landed on two engines with the undercarriage normally extended. Captain R.E Foote and his crew of three were uninjured.

That laconic account conceals a terrifying emergency most efficiently and courageously coped with. Foote had just said that he thought one of the engines was running rough when there were two distinct bangs and the aircraft swung to the right as both starboard propellers disappeared.

The York slowly lost height. All freight, mail, tools and loose gear was thrown out. Harry Foote refused to jettison his suitcase, remarking that he always liked to change after a flight, particularly in the tropics. With no starboard engines and half the starboard elevator torn off, the York just managed to limp into Rangoon.

Foote was now given a commendation from BOAC and went on to fly Britannias over the Atlantic.

But the Rome accident still weighed heavily on him. The International Federation of Airline Pilots, led by Captain Jackson, was still pursuing the case, though they were meeting stone walls wherever they turned.

The authorities refused to release publicly the report on Pentland's Karachi accident. After further performance tests, De Havilland refused to release the stalling speeds of the Comet IA aircraft as applicable to the Karachi take-off accident, either to the Ministry or to IFALPA.

There was now very considerable sympathy with Foote within the Ministry, and at least one official declared that the inquiry should be reopened. But the Ministry clung to a ruling, that an inquiry by an inspector (as on the Foote accident) could not legally be reopened. There was no foundation for this, but it went unchallenged.

The new Chief Inspector of Accidents wrote to Jackson:

I regret to have to tell you that De Havilland's have reached the conclusion that release of the document concerned would tend to mislead since some of the detailed information is open to misinterpretation if taken out of the context of the report, while the whole report is open to misunderstanding if separated from the discussions and actions which preceded and followed it. Under these circumstances they are not prepared to release the report to your Association.

Jackson publicized the letter, adding that IFALPA would call a worldwide strike of its 15,000 pilot members every 3 March (the date of the Karachi accident) unless the information was released.

As a result, some information was released, and a joint IFALPA–De Havilland statement was publicly issued. Guarded and circumspect, De Havilland nevertheless did admit that special tests subsequent to the Karachi crash 'revealed a hitherto unrecognized feature that the stalling speed near the ground was higher than the corresponding figure in free air and that the disparity increased as the aircraft weight increased. The safety margin at the highest take-off weights were thus found to be smaller than that indicated by the certification flight trials.'

Foote was disappointed. He was even more disappointed to receive a letter from BALPA, saying the executive council had approved the statement and the association would take 'no further action in connection with the Rome accident, although it could not hold itself responsible for any action that might be initiated by any individual who had been involved in that accident.'

The *Express* headlined a paragraph on the statement, '*A six year fight clears dead pilot*,' but admitted that it was neither one-sided nor clear-cut. It said pilots throughout the world are expected to be satisfied that Captain Pentland and Captain Foote were wrongly blamed.

Foote was not satisfied. And Pentland was dead. Neither had been cleared. Neither inquiry was reopened.

Foote was again not well. After leaving the Britannia fleet and commencing the conversion course for VC10s, he was again placed in a stressful situation involving hard studying and flying pure jets again. The strain of many years began to tell. He had bouts of breathlessness and angina pains. He had actually completed his VC10 conversion course successfully, but before he could commence route flying, his medical came up. Although he knew the consequences, he told the medical officer about his symptoms and inevitably, after further tests, he lost his licence.

And of course because of the 'anxiety' illness he had had, it was not insured. He had to give up airline flying and tried various jobs on the ground, eventually becoming a simulator instructor.

He and I used to meet with our wives at the White Hart, Lewes, for dinner. For most of the evening, he was his usual amiable quiet self. Then right at the end of the meal, over the coffee, he would bring the conversation round to 26 October 1952. The tablecloth would become Ciampino airport, the guttering candles the runway lights. A knife would be produced, angled to show take-off attitude. Then he would talk into the night about the accident.

Years went by, but he could never forget. He was obsessed with the fact that he had been unjustly blamed and no one would listen.

The last cutting in his fourth scrap-book is headed 31 January 1970 and is from *The Times*. It reports the memorial service for Michael Majendie, and there follows a long list of notables.

Not long afterwards the pains in Harry Foote's heart increased. One afternoon, his wife left the house to visit her mother. Harry Foote told her he thought he would have a bath. When she came home, she called but there was no answer. Going upstairs, she found the bathroom locked. Helplessly she banged on the door, then went downstairs, got a chisel and broke the lock.

Harry was lying face down on the floor. Meticulous as ever, his clothes were neatly folded on the chair, his shoes side by side. In vain, she tried to revive him, and then dialled 999. Within minutes, the ambulance had arrived.They were very gentle.

The obituary notice in the *Daily Telegraph* of 19 August 1970 read, 'Captain Robert Foote of Worthing, aged 53, pilot of the BOAC Comet which crash-landed at Ciampino Airport, Rome in 1952, slightly injuring one passenger.'

There was a simple funeral at Slough two days later. BOAC asked Chris Foote if she wished to have pall-bearers and a company flag. She refused, but she was gratified to find that quite a number of Harry's BOAC and BEA colleagues attended.

There was no public memorial service for Harry Foote. His memorial is now carried in jet aircraft throughout the world – the lights, the bells, the horns, the stick-shakers of the stall warning devices.

In 1981 the last Comet IV was honourably withdrawn from service, and disappeared from the skies. Technically the take-off accidents and the structural failures had been solved by the advance of science.

Why was there what BALPA called 'an aura of secrecy' over the Comet crashes? Why didn't the authorities believe Foote? And why was he never cleared of pilot error? And why did he apparently make a mistake that a boy on his first solo was unlikely to make?

In spite of the 'sense' of not getting the nose too high, which Harry Foote and Charlie Pentland certainly shared with other pilots, the skid marks on the runway certainly showed that they *had* done.

The reason was that the 'sense' depended on being able to see the horizon. Furthermore there was no natural link in the controls. An artificial 'feel' mechanism had been incorporated. When the aircraft was light and the air at reasonable temperature, the angle to which the wing could be lifted safely was a fairly wide one – there was margin for error. But when the aircraft was heavy and it was hotter, it was dicovered *after* the Rome and Karachi crashes, the angle became critical and there was little or no margin for error. Yet in the darkness, there was no yardstick horizon for the pilot to gauge that angle. Foote had naturally not been able to get it right, and the wing appeared to stall, after which recovery was not possible. The same thing appeared to have happened at Karachi.

Numerous experiments in psychology, particularly those done by Ames and Witkin, have shown how helpless man is in judging horizontals and verticals without a horizon. In darkness, people attached to a chair are asked about their posture, whether they are leaning left or right or back or forward. Most get it wrong. Asked to turn a long movable needle to the vertical in darkness, again errors can be large. Certainly there is an artificial horizon instrument on an aircraft, but the nose-up indication is not sufficient in conditions of small error tolerance.

Harry Foote once said that he landed and took off at the exact speeds laid down. He maintained that very few pilots were as accurate as he was, and he based this statement on his experience as a training captain. He had an obsession with accuracy. When he and his wife were on long trips through France and Chris was driving while he rested, he would tell her the exact speed she should drive and would point out to her when the speedometer needle had fallen below or risen above the speed. Pentland was also a training captain and had the same obsession with accuracy. If you have to transmit knowledge to people, you first show by example that you do it exactly right. Had Foote been less accurate, he might have had a few more knots on his airspeed indicator that dark night at Rome.

But then why did it happen only to Foote and Pentland? Partly because of that necessary combination of high weight, warm temperature and darkness had been rare. Partly because many pilots had realized or learned, despite flying instructions they had received, to leave the nose-wheel on the ground until the Comet was just about to become airborne. The wear and tear on the nose-wheel and the noise were considerable, but by that technique of not having even to try to gauge the wing angle, they would not get into a 'ground-stall'. And then of course the formal flying instructions were changed to this technique.

Like Parke before him with spinning, the methodical Foote had stumbled and fallen at a new hurdle of aviation that was not then understood – and had warned of the danger before there were casualties. Had he been believed in time, the Karachi crash would almost certainly never have occurred.

Foote wrote, 'A pilot, not in a position to defend himself, has been used as a public scapegoat. This had happened again at Karachi since the accident in question and will no doubt continue . . .'

Indeed it did continue on frequent occasions – and by no means only in aviation. It is as though society needs such victims. Like others before and after him, he gave his warning. Within the frame of his destiny, he did what he could for us all.

8

The Frame
of Our Destiny

'We are not permitted to choose the frame of our destiny. But what we put into it is ours.'

So wrote Dag Hammarskjöld, a man of destiny himself, who recognized his own role was as vital to peace as Churchill's had been to the free world at war. In 1953, when he took office as Secretary-General of the United Nations, he concluded his inaugural speech to the assembly with the words, 'The greatest prayer for man is not for victory but for peace.'

He initiated the United Nations Meditation Room which is off the public lobby of the General Assembly hall. The only decoration is a block of Swedish iron ore in the middle of the room. Light falls in shafts on it from the windows, and it is dedicated to 'the God whom men worship under many names and in many forms'. He wrote, 'We have all within us a centre of stillness surrounded by quiet.'

Yet though this intellectual Swede, this brilliant economist and statesman, had striven for peace throughout his life, the period of his United Nations incumbency was fraught with conflict. He had sent an emergency force to Egypt at the time of Suez. The beginning of the 1960s saw Africa in tumult. At that time, much of the trouble centred on the Congo, hastily given independence by the Belgians after massive riots at Leopoldville (now called Kinshasa). The copper-rich province of Katanga, led by General Moise Tshombe, broke away from the Central African Federation – thereby depriving the central government of most of its revenue. Fierce fighting broke out aided by the mercenaries of several interested powers.

President Adoula of the Central African Federation and his Prime Minister Patrice Lumumba appealed to the United Nations. Russia also was appealed to.

Dag Hammarskjöld despatched a military force of 20,000 men, mostly composed of Africans and Asians. Their role was primarily to round up the foreign mercenaries. Hammarskjöld became the object of bitter vituperation from the Soviet government, while at the same time, he was uncertain of the support of the West.

In February 1961 Katanga tribesmen captured Prime Minister Lumumba, and eventually, after he had been reputedly tortured by Tshombe, murdered him.

The United Nations force was strengthened. Ethiopia and Eire sent troops. Conor Cruise O'Brien was appointed to the United Nations Secretariat with a special Katanga assignment.

Dag Hammarskjöld was aware that the United Nations was at a turning-point. Drafting a speech he hoped to deliver some months later, he wrote that member nations must choose between it 'as a static conference machinery' or as a 'dynamic instrument' eventually to shape a world community.

The situation in the Congo worsened. The role of the United Nations forces was interpreted differently by Conor Cruise O'Brien and Dag Hammarskjöld. United Nations forces entered Katanga. The non-violent UN intervention deteriorated into fighting. President Adoula asked Hammarskjöld to intercede and arrange a meeting between General Tshombe and himself.

In his diary, Hammarskjöld wrote, 'Once I answered Yes to Someone – or Something. And from that hour, I was certain that existence is meaningful and that, therefore, my life in self-surrender has a goal.'

In acceding to Adoula's request, Hammarskjöld recognized that the meeting would be both difficult to arrange and dangerous. Too many interests wanted the war to continue, too many of the most dangerous foreign mercenaries were still at large.

Hammarskjöld arrived at Leopoldville to find an even more confused situation. Though President Adoula trusted his judgment and integrity, many Africans were suspicious. The Soviet government continued its opposition, the Western powers dragged their feet. To add to the confusion, Lord Lansdowne, the Under-Secretary of State at the British Foreign Office, appeared on the scene. His appearance gave rise to rumours that the West were trying to put pressure on Hammarskjöld, thereby making any move he made that much more difficult. Western interests were known to favour Tshombe and Katanga. And so mistrustful was the Congo central government of Lord Lansdowne that they considered expelling him.

Lansdowne's visit coincided with reportedly heavy fighting and some Katanga victories. Elisabethville (today called Lubumbashi) was in United Nations hands, but in Jadotville, seventy-five miles to the north-west, United Nations troops were overwhelmed by Katangans, and an Indian armoured column marching to relieve the United Nations forces were strafed by a Katanga Fouga fighter, reportedly flown by a Rhodesian pilot.

The Potez Fouga was a light trainer powered by two turbojet engines, built for the French Air Force. The pupil sat at the front, the instructor in the rear. Germany bought 250, and on 15 February 1961 three armed Fougas were delivered to Katanga. Four 25-kilo air-to-ground rockets or one Nord guided missile could be fitted under each wing. Maximum speed at sea level was 441 mph – far faster than the DC6 – and its range was 735 miles.

Tshombe did not immediately commit himself to the meeting. Hammarskjöld's plan was that they should meet at Ndola in Northern Rhodesia, and if the general proved amenable, he would accompany him to meet President Adoula. And while these delicate negotiations were going on, the United-Nations-controlled radio station in Elisabethville put out a message saying it was determined to end the secession of Katanga, and that civilians caught as spies would be shot.

Hammarskjöld thereupon received an arrogant evasive reply from Tshombe. He might meet Hammarskjöld, but he might not. Finally another message came to the effect that Tshombe's plane was at the end of the airstrip runway at Kipushi about to take off for Ndola.

Hammarskjöld immediately made arrangements to fly to Ndola to meet him. Normally he would have used the Transair DC4 as he usually did. But this aircraft was to be used by Lord Lansdowne. Instead another Transair plane operating under charter to the United Nations, a DC6B, SE-4BDY, was flown from Elisabethville to pick up Hammarskjöld and his party at Leopoldville.

The situation in Elisabethville was still tense. Shots were fired at the DC6 as it took off. On arrival at Leopoldville on the morning of 17 September, it was discovered that a bullet had shattered the exhaust pipe to Number 2 engine. The damage was repaired. Routine pre-flight checks were carried out. The fuel and oil tanks were refilled.

Then the DC6 was allowed to stand on the airfield unguarded for between three and four hours – a strange oversight in view of the complicated, somewhat cloak-and-dagger precautions that were taken about the actual flight.

This flight was commanded by Captain Hallonquist, an experienced Swedish pilot and an expert in navigation. On the suggestion of the Leopoldville air traffic control officer, Captain Hallonquist filed a departure from Leopoldville to Luluabourg, 400 miles to the west, and not to his real destination, Ndola.

To add to the confusion, the DC4 OO-RIC, which had been loaned to Lord Lansdowne, did leave Leopoldville for Ndola shortly before Hammarskjöld's DC6. There are two main versions as to why. One version is that Lord Lansdowne was going ahead of the Secretary-General to smooth the arrangements between him and Tshombe, and to discuss matters with Lord Alport, the British High Commissioner in Rhodesia, who was also now in Ndola.

To this was added the explanation accepted by the Rhodesian and Nyasaland accident inquiry, that the old DC4 with Lord Lansdowne aboard went ahead to act as a decoy for the DC6 and Dag Hammarskjöld. It seems strange that Lord Lansdowne, who had reportedly earlier that day informed Hammarskjöld of British dismay, and who had shown a marked lack of enthusiasm for allowing United Nations fighters to overfly British territory, should now become Hammarskjöld's decoy.

Whatever the object of Lord Lansdowne's flight to Ndola, his DC4 was airborne from Leopoldville at 15.04. It headed south-east. Forty-seven minutes later, Captain Hallonquist lifted the DC6 off, and though bound for the same destination, headed west by a much longer and more complicated route. The purpose of this circuitous route was to avoid attacks from the ground and from the Fouga fighter.

Radio silence was maintained until the aircraft was over Tanganyika territory, four hours after take-off. As instructed, after that all messages were sent in morse and Swedish.

Yet in spite of such complicated security measures, the Captain had been given no briefing for the flight while he was at Leopoldville. He was reported to have told only one person, a United Nations major, his true destination.

Meanwhile the DC4, OO-RIC with Lord Lansdowne on board, having flown

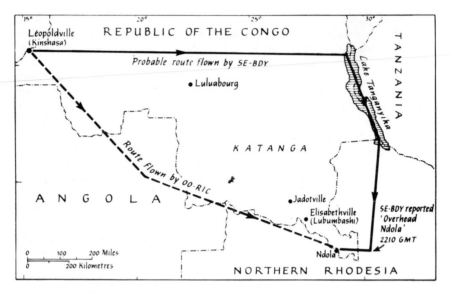

the direct route in full radio contact with navigation lights on, arrived at Ndola at 20.35 GMT without incident.

At 20.40 SE-BDY reported to Salisbury flight information region, 'Flying above Lake Tanganyika at 17,500 feet.' Sixteen minutes later, it was abeam Kasama, estimating Ndola at 21.47.

Ndola air traffic control sent the weather: 'Surface wind 12° magnetic, speed 7 knots, QNH [sea level pressure] 1021 millibars, QFE [aerodrome pressure] 877 millibars, visibility 5–10 miles with slight haze.'

At 21.38 Ndola cleared the aircraft to 6000 feet.

SE-BDY checked in abeam Ndola at 21.47 and at 22.10 reported, 'Lights in sight. Overhead Ndola, descending. Confirm QNH.'

Ndola replied, 'QNH confirmed 1021 millibars. Report reaching 6000.'

SE-BDY did not report at 6000 feet and no further communication was received from the aircraft.

Eyewitnesses reported it flying low nearly a mile to the north of the airport beacon and continuing above a house seven miles to the north-west. The correct instrument procedure stipulates a let-down to the *west*.

Meanwhile the DC4, OO-RIC, with Lord Lansdowne on board was waiting on Ndola runway. When SE-BDY did not land, it took off. Lord Lansdowne is variously reported as returning to Leopoldville and flying to Salisbury. When Hammarskjöld's aircraft still did not land, for some unknown reason it was considered simply to have flown off to another destination. And in spite of the police reporting fire in the bush about ten miles from the airport, no emergency was declared and no search was made.

The Times reported that Sir Roy Welensky, the Rhodesian Prime Minister, promised mercy help to Katanga, and when asked how supplies to opponents of the United Nations placed Rhodesia he replied, 'I don't know how it places us. I don't really care. If I get a request from starving people, I am prepared to help them.'

Next day natives were discovered selling pieces of aircraft and blackened clothes in a local market. A light aeroplane was sent up and the wreckage of SE-BDY was immediately spotted in line with the runway heading of 100° true and nine and a half miles short of the threshold.

It had hit trees at an altitude of 4357 feet at a normal landing approach angle at a normal approach speed. It had then struck an ant-hill, cart-wheeled and caught fire.

The undercarriage was down and locked. Thirty per cent of flap had been selected. The engines were under power and were fully serviceable. The altimeters had been set correctly on the setting given by the Ndola controller. There was no sign of the aircraft having been hit by bullets. One examination that was not made was the remelting of all the fused metal recovered from the point where fire had occurred, in order to see whether any projectile could be found. The metal had all been broken by hammer and steam-hammer into pieces about eight inches square and a few inches thick. It was decided that the melting of all this wreckage was not justified.

There was only one survivor, an American. He spoke incoherently about 'great speed' and said Hammarskjöld 'changed his mind and said "turn back"'. Since the aircraft was just coming in to land, that was considered unlikely. The man died shortly afterwards, still very incoherent.

One puzzling detail was that the bodies in the cabin were riddled with bullets. This naturally gave rise to a story that the aircraft had been ambushed. Later examination by pathologists and gun experts indicated that the bullets had not been fired. The heat of the fire had exploded them on the bandoliers (shoulder belts of ammunition) worn by Hammarskjöld's body guards, sending them into the bodies of those opposite.

The bodies were laid in a banana store, in spite of the owner's fears that his customers would hear of it and desert him. Dag Hammarskjöld had been thrown clear and was not burned. He probably was alive after the crash and might well have survived had help come earlier.

Found beside him was Buber's *Ich und Du*, which Hammarskjöld had taken with him on the flight. Left behind was Thomas à Kempis' *Imitation of Christ*, inside which was found a copy of the United Nations oath of office.

There being no Swedish flags in Ndola, they were run up by a local tailor with the guidance of an illustration in a school atlas. These home-made efforts covered the coffins of Hammarskjöld and his Swedish compatriots. Tshombe laid a wreath on the Secretary-General's coffin which was then flown back to Uppsala for a state funeral.

There was a wave of anti-European feeling, as Tshombe was suspected of being the tool of imperialists. Joshua Nkomo called the crash 'a serious indictment of the British government'. The *Indian Express* wrote, 'Never even during Suez have Britain's hands been so blood-stained as they are now.'

Shortly after his appointment as Secretary-General, Hammarskjöld wrote for a programme of Ed Morrow's about his source of inspiration in the

writings of those great medieval mystics for whom 'self-surrender' had been the way to self-realization, and who in 'singleness of mind' and 'inwardness' had found strength to say Yes to every demand which the needs of their neighbours made them face, and to say Yes also to every fate life had in store for them.

On 23 October 1961 the Norwegian Storting awarded the Nobel Peace Prize to Dag Hammarskjöld.

There were three inquiries into the accident. The Swedish inquiry found no evidence of pilot error. The one carried out by the Federation of Rhodesia and Nyasaland found no indication of mechanical failure, defective altimeters, internal fire, incapacitation of the pilots, sabotage, being shot down from the ground or by another aircraft.

It had been a long and stressful day for the crew, and the co-pilot had complained of feeling tired when he boarded the aircraft at Leopoldville. But the inquiry did not consider that fatigue contributed to the crash, nor did it find evidence that incorrect information was given to the aircraft, either by Salisbury or Ndola.

Every airliner carries a manual in which are the approach charts and let-down procedures of the various airports. Their height above sea level, the high ground and obstructions and the arrangement and length of the runways are all marked. On preparing to land, the chart of the airfield being approached is conveniently positioned between the two pilots so that *both* of them can see it. No matter how good the weather, no captain would attempt an approach at night to an unfamiliar airport without the chart except for a very pressing reason. Captain Hallonquist had never landed at Ndola before.

Pilots in Transair received individual issues of the Jeppesen's manual of approach charts, including one for Ndola. One of these manuals was found in the wreckage. The Ndola chart was missing.

Three other manuals of approach charts were also on board SE-BDY. These were bound volumes of United States Air Force approach charts, and although Ndola had been an airport for years, the 1961 issue contained no approach chart for it. But it did contain a chart for Ndolo, an airfield six miles from Ndjili airport at Leopoldville. One of these USAAF approach manuals was found in the wreckage – opened and folded back to show the Ndolo chart.

The height of Ndola above sea level is 4160 feet. The height of Ndolo is 951 feet.

The inquiry considered that the aircraft was being brought in by a visual descent. It was a clear night, all the airport lights were at maximum intensity, SE-BDY had reported seeing them and had been told there was no other traffic.

There was a slight mist on the approach to the airport. It was thought that this might obscure the runway lights in the course of a low turn and the Rhodesian and Nyasaland inquiry – in spite of being strongly urged not to bring in a verdict of pilot error – 'felt it must conclude that the aircraft was allowed by the pilots to descend too low. In so doing, it struck trees and crashed'.

The possibility that the captain let down on the Ndolo chart, with its height given as 951 feet above sea level, thinking it was the Ndola chart (height 4160 feet), was strongly rejected.

The inquiry's theory (totally excluding the possible contribution of mistaking Ndolo for Ndola in the manual) would have been more credible had SE-BDY crashed on a steeply descending turn into the approach. But the wreckage was reported dead on line with the runway heading on a normal landing descent angle. In this position, the pilots were bound to have seen the runway lights. But judging height or distance from lights is notoriously difficult, and both pilots would be busy with landing checks. It is unlikely that an experienced pilot would undershoot in clear weather by nine and a half miles, unless he was under the impression that he was far higher than he was. Had he been using the Ndolo height of 951 feet above sea level as the height of Ndola, he would think he had several thousand feet still to spare, which he would proceed to lose at the normal landing descent angle found on the plane after it had crashed.

Then why was the Ndola approach chart missing from the manual? And why, when SE-BDY, with such an important passenger on such an important mission, failed to land from its approach, was not an immediate search and rescue operation initiated? That would have been undertaken automatically as a matter of course, without the fact of a police-reported fire to spur it on. The authorities believed that SE-BDY had simply gone back to Leopoldville. But why would it? And would it have enough petrol? The whole purpose of its trip had been to enable Hammarskjöld to meet Tshombe. The weather was good. Why would all the elaborate and difficult efforts to effect the meeting be thrown away? Why, if the aircraft *did* divert, had not the captain let Ndola know? The theory that he was preserving radio silence is not tenable, since he had broken it as soon as he flew out of Congo territory and he had continued to break it over Ndola on his landing exchange. The impression left by the lack of follow-up is that for some reason someone knew that the captain would have a problem in landing, and that the meeting between Tshombe waiting on the ground at Ndola and Hammarskjöld in the air above the airport would never take place.

The inquiry conducted by the United Nations Organization did not throw any further light on the mystery. Its report regretted that no information had been given to the responsible official of the United Nations Organization in the Congo (UNOC) of the route which the Captain intended to follow after take-off from Leopoldville. Another report stated that in fact he told a UNOC major. The UN also regretted that no security measures had been taken to guard SE-BDY while it was on the ground.

The inquiry considered the possibilities of material or instrument failure, fire on flight, pilot incapacitation and pilot error, and attack from the air or ground. No evidence of any of them was found, but neither could the possibilities of any of them be excluded.

With respect to sabotage, the report stated that

the aircraft was without special guard whilst it was at Ndjili airport in Leopoldville and access to it was not impossible. The Commission is aware that there are many possible methods of sabotage. No evidence of sabotage has come to its attention but the possibility cannot be excluded.

The United Nations troops left the Congo at the end of June 1964. Zaire came into being in October following the dissolution of the independent Federation of Rhodesia and Nyasaland.

Now more than twenty years later has come, not actual evidence, but a very positive assertion of sabotage, an accusation directly levelled at General Tshombe. Three years after Hammarskjöld's death, Tshombe had succeeded Adoula as President of Zaire. His tenure was brief. He in turn was overthrown in October 1965 and fled to Madrid. He was sentenced to death in his absence. It was suspected that while in exile he planned another coup,

which certain European interests might support. Once again there was plotting
and counter-plotting. A Frenchman named Francis Bodenan succeeded in
penetrating Tshombe's entourage in Madrid and in winning his confidence. On
30 June 1967 Tshombe and his henchmen, including Francis Bodenan, boarded
a Hawker Siddeley 125 aircraft of Gregory Air Service bound from Ibiza to Palma
Majorca. Over the Mediterranean Bodenan hijacked the aircraft and forced it
to land at a military airfield near Algiers where Tshombe was arrested,
imprisoned and two years later died of, it was said, a heart attack.

Francis Bodenan was arrested and began to talk. He was acting, he said, as
an agent for Zaire, to prevent Tshombe causing another uprising. He claimed
Belgian and Spanish connivance in his hijacking. He said that he had turned
against Tshombe after seeing evidence of Tshombe's involvement in the death
of Hammarskjöld. Bodenan named a Belgian who had shown him a 120-page
report on Tshombe's part in the killing of Lumumba and the Hammarskjöld
aircraft crash.

Furthermore, Bodenan, who has recently been tried by a Spanish military
court, declared that in 1967 Tshombe himself told him exactly how the
Hammarskjöld crash had been brought about. Posing as a technician, one of
Tshombe's secret agents boarded the DC6 while she was left unguarded on
Ndjili airport, Leopoldville.The secret agent removed the map carrying the
flight information about the landing field at Ndola – the map in fact whose
absence had been so puzzling. Then when the pilot approached Ndola and
radioed for a position check, the traffic controller there allegedly gave him
incorrect information.

Though murder and sabotage were rumoured at the time of the crash,
though Tshombe with his history of revolution and butchery was a prime
suspect, not until March 1982 were such assertions actually made. Does this
testimony solve the mystery? Or is it only, as *The Times* headlines stated, that
the '20 year mystery' deepens?

Bodenan's explanation would certainly fit the facts both technical and
political. Many, Tshombe included, found Hammarskjöld's efforts for peace
unwelcome. Was Tshombe helped by other interested parties? For this 'truly
good man – his integrity was absolute', as President Mobutu called
Hammarskjöld, was the subject of attack from the Russians, some of the other
African countries, and the Katanga copper lobby in the West. Yet Mobutu
declared in 1976, 'The Democratic Republic of the Congo is a living testimony
to what the United Nations is capable of when it is given the appropriate
means. In the Congo, the United Nations defended the sovereignty of a country
which certain covetous interests were ready to compromise.'

He also said, 'He was a man involved in momentous issues which he came to
symbolize even in death.'

Hammarskjöld's death did not impede his work. The shock, the suspicions,
the anger it evoked enhanced it.

Was his death necessary? Was he a man of destiny? Or was the crash caused
by a number of factors, including the pilot's fatigue? Was it sabotage from the
ground? Or was it Tshombe's plot – the simple deadly expedient of the map
extracted from the unguarded aircraft?

Ironically, the very last entry in Hammarskjöld's personal notebook is a poem. It begins, 'Is it another country/In another world of reality?' And it ends, 'And I begin to know the map/And to get my bearings.'

Men of destiny bear certain resemblances to each other. Frequently they are inspired by a powerful dream or impelled towards a sometimes self-destructive goal. Hammarskjöld's goal was world peace. General Sikorski's was a free Poland. Both were fatal.

It has been said that the tragedy of Poland was written on Sikorski's face. Born in Galicia in 1881, he was the son of a gentleman farmer who died when he was a child. He studied at Cracow and Lvov and took a brilliant degree in engineering. During the first war, Poland was divided within itself – part Austrian-dominated, part Russian – and in 1920 Sikorski fought a Russian offensive to the gates of Warsaw.

A year later, he was Chief of the General Staff and in 1922 became Prime Minister. In 1926, after Pilsudski's *coup d'état*, he retired to Paris for ten years to think and write. On the eve of the Second World War, he took command of a volunteer force of Polish miners and workers. They fought magnificently till the capitulation of France, after which he and 24,000 Poles were evacuated to Britain where he declared they would continue the fight. A few days before his death, he spoke proudly of the courage of his men and said, 'I am glad to tell you that I shall myself be in command of the first troops to enter Poland.'

Britain and France had entered the Second World War against Germany to preserve Poland. When Hitler turned on Russia in 1941 conversations began between Sikorski, as Prime Minister of the exiled Polish government, and the Soviet Ambassador, M. Maisky, under British auspices. Sikorski wanted the recognition by the Soviet government that the partition of Poland agreed by Germany and Russia in 1939 was now nullified, and the liberation of all Polish prisoners deported to the Soviet Union after the Russian occupation.

Russia would agree to neither. The hard-pressed British, desperately anxious for Russian support, were in a fix. Churchill had no alternative but to ask Sikorski to postpone his perfectly natural and legitimate demands till after the war against Germany was won. After great bitterness, a Soviet-Polish agreement was signed, the British putting on record that they did not recognize any territorial changes which had taken place since August 1939. At the same time, they did not guarantee any frontiers.

It was perfectly clear that after helping to win the war, Sikorski and the Poles were going to pose an insoluble problem.

Strange things had started to happen when Sikorski was on a flight over the Atlantic. Two hours out a stick bomb was discovered in the Elsan toilet of a British Liberator in which he was travelling to America. According to MI5, however, this bomb had come into the possession of a Polish wing-commander on Sikorski's staff. For reasons of his own, this man had kept it quietly in his brief-case till the whines and bangs of the Liberator battling against the Atlantic gales had convinced him that it was going to go off. He rushed it to the toilet and put it down. He then apparently discovered that the Elsan had inadequate bomb-quenching properties and immediately reported his 'find' to a fellow passenger. Everyone assumed that an attempt had been made on the

General's life. Further facts were eventually discovered, and the matter dealt with by the Polish authorities.

Not long afterwards, both engines of an aircraft in which Sikorski was travelling failed on take-off from Montreal. General de Gaulle, after similar escapades, had stopped travelling by aeroplane.

Meanwhile relations between the Poles and the Russians had worsened. After the Russo-German agreement of 1939 to partition Poland, 14,500 Poles, mainly officers, were put into three camps round Smolensk. Nothing had been heard of them since April 1940. Naturally, now they were allies of Russia, the Poles wanted these 14,500 released to join in the united attack against Germany. Nobody was released. No news at all was forthcoming.

In April 1943 Sikorski told Churchill over lunch at 10 Downing Street that he had proof that the Russians had murdered the 14,500 and buried them in vast graves in forests round Katyn. The Poles asked the International Red Cross in Geneva to conduct an on-the-spot inquiry. Shortly afterwards, German radio charged the Russians with the murder of the 14, 500 Poles in the three camps, and proposed to hold their own on-the-spot inquiry. The Red Cross, meanwhile, had said they could not conduct an inquiry without an invitation from the Soviet Government. So the Germans alone produced a detailed report, claiming the discovery of over 10,000 bodies in mass graves in the Katyn area.

The Poles demanded an inquiry. Maisky told Churchill that on account of Polish accusations the Soviet-Polish agreement of 1941 was at an end.

The Polish forces had fought gallantly for the Allied cause. But the Allies did not appear to be fighting for a free Poland, and one of them, Russia, did not even recognize the Polish government in exile.

Naturally Polish morale was low. In May 1943 Sikorski had flown out to the Middle East to raise their spirits. Two months later he set off again.

He arrived in Gibraltar on 3 July on his way to the Middle East with his daughter, his Chief of Staff, Colonel Victor Cazalet and others in a Liberator piloted by a Czech pilot attached to the RAF Transport Command, called Prchal, with 600 hours as a Liberator pilot. Sikorski had flown with Prchal several times and had specifically asked for him.

Gibraltar's one runway had been lengthened since I was there, but was still very short. Then we had waited an extra day for the wind to get up before taking off our aerial-decked radar Wellingtons. Even then, some doubt had been expressed about our getting airborne. Dangerous air currents were supposed to lurk round the high rock that overshadowed the runway to the south. A few yards to the north, beyond the barbed wire in Spain, Axis observers checked every movement and radioed the news ahead into the German-controlled Mediterranean.

The sea at both ends of the runway was the graveyard of many aircraft.

Throughout 4 July Sikorski's aircraft had stood on the tarmac, heavily guarded. By an odd coincidence, Maisky happened to be in Gibraltar at the same time.

That night General Mason Macfarlane, Governor of Gibraltar, accompanied Sikorski and his party to the airfield and said goodbye. Then with others he watched Prchal taxi the Liberator to the end of the runway.

The Czech pilot opened up the throttles and began roaring down the runway. The opinion has since been expressed in books that Prchal had a 'trademark' of 'hauling a Liberator off the ground' and then putting the nose down to gain a large safety margin of speed before beginning to climb up and away – and that such a technique was used that night, resulting in the plane becoming airborne 500 yards before the end of the runway.

Perhaps the reporting has been wrong, but I have flown 3500 hours as a captain in Liberators and have never heard of such a technique. Because of its small wing, a heavily laden Liberator's take-off was notorious. You normally let it use the whole of the runway before pulling it off and hoping for the best. There was no real question of holding it down before climbing. You stayed low because you could not climb till you had notched up a few more knots. To pull off any aircraft too soon is dangerous.

Shortly after becoming airborne and crossing the runway threshold, to the horror of those watching, the Liberator lost height and plunged into the sea.

The airport searchlights focussed on the wreck before it sank as launches rushed to the rescue.

Prchal, the captain, was rescued with only minor injuries. General Sikorski, his daughter, his Chief of Staff, Victor Cazalet and everyone else on board were killed.

The Germans immediately gave out on the radio that the Russians had assassinated Sikorski. The Russians accused the British. Stalin was reported to have said the British had put a secret agent on board the Liberator. Maisky considered he had been hurried away from Gibraltar so that the British could kill off their inconvenient ally quietly.

Prchal said that he had tried to climb but the controls appeared to be locked solid. He could do nothing else, he said, but to throttle back and try to ditch ahead. An Air Ministry communiqué accepted this as the reason for the crash and absolved Prchal from blame.

Subsequent investigation revealed no evidence of control locking. There were several inquiries, none of which threw any light on what had happened.

The cause of the crash remained a mystery. There were several other interconnected mysteries. Six weeks before the crash, three high-ranking Polish officers had received phone calls to say that Sikorski's plane had crashed at Gibraltar and all its passengers had been killed.

Prchal had taken off as usual, it was reported, with his life-jacket draped behind his seat. Yet when he was rescued from the sea, he had the Mae West on with every tape and fastening done up.

The bodies of the victims were to be taken back to Britain in the Polish destroyer *Orkun*. But the captain refused to take them, saying it would spell doom to his ship to turn it into a hearse. He was overruled on the Poles, but the British Victor Cazalet had to be buried at Gibraltar. Three months later the *Orkun* was torpedoed and sunk.

At that same time Katyn was reoccupied by the Russians, who conducted their own inquiry into the alleged massacre. They maintained that the three camps had fallen to the advancing Germans who had themselves massacred the 14,500 Poles. It is significant that at the Nuremberg inquiry after the war Katyn was skated over and never mentioned. A conspiracy of silence surrounded the whole affair.

Years later the mystery flared up again when the German Rolf Hochhuth wrote a play called *Soldiers*, inspired by Churchill's phrase, 'Soldiers must die, but by their death they nourish the nation which gave them birth.' Hochhuth's play appeared to be based on the idea that Churchill had ordered the assassination of Sikorski as a prelude to the mass bombing of German cities. How this was accomplished is not revealed – but *that* was put forward as the cause of the accident on 4 July 1943.

Gibraltar was a dangerous airfield. The only runway was too short. It was found that Sikorski's aircraft was tail-heavy, that extra liquor had been taken on, unknown to the pilot. But a heavily laden Liberator had a notoriously long and difficult take-off. I have known of too many Liberators which crashed on take-off to need to look for any other reasons. Most pilots have a sixth sense not to pull the Liberator off too soon.

9

The
Time Machine

In the faded photograph of 1907, they look a most unlikely trio to be Men of Destiny. And yet in their time they were. For those three were the founder members of the Royal Aircraft Establishment at Farnborough, Britain's world-famous aeronautical engineering establishment. And their goal was to conquer the sky.

In the centre, Colonel Capper, the director – moustached, uniform hat as straight as a die, renowned for setting up courses 180° out on compasses. On the right, Samuel Franklin Cody, an illiterate American genius, hair down to his shoulders, pointed waxed moustaches, wearing riding breeches and a ten-gallon cowboy hat, first to fly an aeroplane in Britain in 1908. On the left, dapper in straw hat and lightweight suit, Lieutenant J.W. Dunne, aircraft designer, aviation prophet, the most important figure – so J.B. Priestley, who was fascinated with the same subject, was to call him – 'in the campaign against the conventional idea of Time'.

Born in 1875, Dunne served in the Imperial Yeomanry during the Boer War, where he became interested in the flight of sea-birds, particularly the albatross, which he sought to emulate with paper models. Watching them sweep their wings back at speed, it seemed to him that the most efficient aerofoil was an arrowhead, and he began making flying models and later monoplanes on that design which he launched from the hill at Farnborough over Laffan's Plain.

In these, with astonishing accuracy, he was forecasting the future. No other early aircraft designers – the Wrights, Cody, Blériot, Sopwith – built other than rectangular wings, and not till the jet fighter of the 1950s is Dunne's bold sweep-back of the leading edge seen. They were not the success they should have been because at that time there was no engine light and powerful enough to propel them efficiently.

Dunne's neat modern monoplanes – against the spindly cobwebs of their contemporaries, they can clearly be seen to be decades too early – were the crystallization of his other great interest, Time. There had been many prophets of flying before him – Leonardo da Vinci, Cayley and Jules Verne. Dunne was particularly inspired by H.G. Wells' writings on both flying and Time, especially *The Time Machine*, in which the hero selects various centuries, both past and future, in which he would like to live and is transported to them, not by magic as would have happened in previous stories, but by Wells' twentieth-century 'scientific mechanism'.

For it was not only the shape of aeroplanes to come that Dunne forecast. He also had the most extraordinary precognitive dreams. Not only in aeroplanes was he finding himself out of step with Time. Though he did not claim to be psychic, in his dreams he began to see things taking place before the date in which they happened.

While he was in South Africa, he dreamed that he was standing on a mountain in the middle of an island, watching smoke rising out of cracks in the rocks. He was filled with the immediate certainty that there was going to be a terrible explosion, that he was about to witness a similar volcanic catastrophe to the one on the island of Krakatoa, twenty years before. In his dream, he pleaded with French authorities to send help, and he woke up shouting, 'Listen! Four thousand people will be killed unless – '

A short time afterwards, Mount Pelée on the island of Martinique erupted with the loss, according to a newspaper, of 40,000 people. The newspaper figure was exaggerated, but made Dunne believe – because of the similarity of the figures and the ease with which a nought could be left out – that he had not 'witnessed' the catastrophe but in his dream had read the newspaper account. John Buchan was to borrow the idea so that a man sees his own obituary in *The Times* of years ahead.

After that he always kept a pencil and paper by his bed. Amongst others he jotted down in considerable detail was a dream of a fire at a rubber factory in Paris the day before it happened, in which girls who had escaped from the burning building were suffocated by the smoke. Another of his dreams was about a runaway horse that had managed to jump a fence and pursue him towards a flight of wooden stairs between narrow railings – which materialized in actuality next day.

If things could be seen before they actually happened, then the accepted idea of Time was upside down, and he began to work out a mathematical concept of how Time could include such anomalies, eventually producing his classic *Experiment with Time* and later *The Serial Universe* and *Nothing Dies*.

According to him, Time is linear and events are spaced along it. Time flows and people experience the events. But since it flows, there must be another Time that regulates the speed it flows, and another Time that regulates that and so on to infinity. This means there are an infinite number of parts of the individual. One part is passive. One part observes a particular way, another a different way. One sees ahead because the events are already there.

Loosely connected to Einstein's Relativity, it is not a particularly satisfactory theory, but it did jerk people out of rigid conventional thinking that Time and the clock were synonymous. In the 1940s the New Look in psychology shattered the previous 'structuralist' ideas that we should perceive what is in front of us if we pay attention. Numerous experiments showed that we see what we want to see, need to see and expect to see, not necessarily what is there. *An Experiment with Time* provided the same impetus to look at Time from another angle, but no theories or experiments so far have shown what it is or how precognition and premonitions can be explained.

From personal experience, I am quite sure that precognition *does* exist. There have been too many strange events in my own life for me to think otherwise. I believe many others have had similar experiences, and there have also been too many cases that are now well authenticated. Somehow, some people *know* certain events are going to happen.

In Somerset Maugham's play *Sheppey*, a man sends his servant into the market at Baghdad to buy provisions. While he is there, he feels a touch and looks into the face of a white-faced woman. She looks at him strangely and makes a gesture. Fearful, the servant runs home and tells his master that Death had threatened him. He asks for a horse so that he can leave Baghdad and go to Samarra, so avoiding Death. His master gives him the horse and watches him gallop away. Then he himself goes to the market and there sees the white-faced woman. He asks her why she made that threatening gesture at his servant. That was no threat, Death replied. That was surpise at seeing him in Baghdad when she had an appointment with him at Samarra.

Numerous early aviators like Sefton Brancker, Amy Johnson and Amelia Earhart were perfectly well aware that they were on the road to Samarra and did nothing to turn off.

In 1884, in an editorial footnote to a story of a sinking liner where most of the passengers were drowned, W.T. Stead, the famous editor, social worker and spiritualist, wrote, 'This is exactly what might take place, and what will take place, if liners are sent to sea short of boats.' Eight years later, he described the sinking of a liner by collision with an iceberg. He said he saw himself dying by 'violence and one of many in a throng'. A medium warned him that his life was 'in danger from water and nothing else', and ten months before the *Titanic* sailed predicted that 'travel would be dangerous to him in the month of April 1912'. But despite numerous such warnings he sailed on the *Titanic* and was drowned.

The invitation to the 'appointment at Samarra' is sometimes foreseen and side-stepped, or foreseen and *cannot* be side-stepped. In 1925 a distinguished test pilot for Fairey Aviation called Captain Norman Macmillan conceived the idea of Britain winning the world's long-distance flying record in a new Fairey aeroplane called the Fox. Dick Fairey was enthusiastic, and plans were made – since there were only grass airfields at that time – for the Fox to take off on the hard tarmac of Eastern Avenue, one of London's new ring-roads.

Then Fairey had doubts about the Fox heavily laden with petrol taking off in a built-up area – and the whole idea was shelved. However, Fairey promised his senior test pilot, 'If we ever build a long-distance aircraft, Mac, you shall fly it for the record.'

Fairey himself was a down-to-earth aircraft manufacturer who considered that everything could be explained and had no time for any ideas on extra-sensory perception. It was in his view simply foresight that two years later he should have been awarded a contract by the RAF to build a special plane to capture the world record.

Naturally Macmillan would be doing the testing. The aircraft was called the Postal Monoplane, and naturally Macmillan was expecting to fly it, as Fairey had promised, for the world long-distance record.

It therefore came as a surprise to him when he had to report to the Air

Member of Supply and Research, Air Marshal Sir John Higgins, otherwise known as Bum-and-Eyeglass,[1] at Air Ministry. There he was asked to release Fairey from his promise, because 'the honour of this flight must belong to the Royal Air Force. It cannot be shared with industry'. It was further intimated that if he refused, the contract would go elsewhere.

Macmillan had no option but to agree. Squadron Leader Jones-Williams and Flight Lieutenant Jenkins, both very experienced RAF pilots, would be flying the machine. When Macmillan told his wife, she said, 'It may all turn out for the best.' Fairey never mentioned the matter.

Meanwhile, the building of the Postal Monoplane proceeded at Fairey's Hayes factory. By the summer of 1928 most of the parts had been made, and its 570 hp Napier Lion engine had been thoroughly tested. Macmillan had a lot to do with arrangements in the cockpit, the siting of the levers and the instruments. Jones-Williams and Jenkins came down to the factory, looked the designs over and expressed themselves satisfied. They never mentioned the Fairey promise to Macmillan.

The parts were then taken by road from the Hayes factory to Northolt, and there assembled, given the RAF identification J9479 and finished in silver dope.

On 30 October 1928 it was passed by Fairey's inspectors and wheeled out for its first flight. Word had got around, and a large crowd assembled at the airfield to watch.

Macmillan climbed on board, tested everything carefully, then slowly taxied forward. Taxiing trials were completed satisfactorily, then fast runs up and down the whole length of the field.

He then put on full power and inched the aircraft just off the ground. This too was a normal prototype test. The Postal Monoplane behaved perfectly satisfactorily, but suddenly Macmillan was seized by an overwhelming sense of foreboding.

He throttled back and came back to earth. There should have been further hops just above the ground, gradually getting longer before the awaited leap into the air. But Macmillan taxied the aircraft back to the hangar. There he said that someone else could take the aircraft off. He would rather not handle it.

He told Fairey, 'This afternoon, I felt I was in a flying coffin. It was a horrible feeling.'

A new type's first flight is an unknown and dangerous business. But it was unheard of for a test pilot of Macmillan's quality and experience to refuse to fly one and for no technical reason. Even so, Fairey did not argue the point one way or another. But a few days later he did ask if Macmillan would change his mind, as the Air Ministry, having heard of his refusal to fly, now had the idea that there was something wrong with the Postal Monoplane.

Macmillan still refused and continued his ordinary test flying duties on other aircraft. While he was down at the firm's Hamble office, the telephone rang and he was told by the works manager that Squadron-Leader Jones-Williams had flown the Postal Monoplane.

He said nothing, but during the next few days felt himself ostracized. There

1 He was also the AMSR for most of the R101 story. See 'Beyond the Inquiry'.

was no doubt that he was considered to have funked it. Even the fact that he now agreed to test fly the aircraft did not eradicate his isolation.

Every time he took the aircraft up, he had that same strange feeling that he was in a flying coffin.

During the test period, he saw Jones-Williams and Jenkins several times. But it was evident they did not want to meet him. A wall had come down between them.

On 24 April 1929 Jones-Williams and Jenkins set off from Cranwell in the Postal Monoplane to try for the record which then stood at 4460 miles. They flew over the Middle East and beyond Karachi. But their fuel was low, and after 50 hours and 38 minutes, they landed still short of the record.

Bum-and-Eyeglass Higgins was at Northolt aerodrome where, after easy stages on the home trip, they landed safely.

'Well,' he said, turning his eyes round to look at Macmillan standing behind him. 'I don't see anything very superstitious about *that!*'

The story had got round and continued to haunt him. He was still being ostracized. Modifications were being made to the Postal Monoplane to give it greater range for another shot at the record, which five months later was broken by the French and now stood at 4912 miles.

After the modifications had been completed and Macmillan had done the test flying, he was asked by Air Chief Marshal Trenchard what he thought of the Postal Monoplane now.

'They have improved her technically,' he replied 'But basically she is unchanged.'

Early in December Jones-Williams and Jenkins arrived at Northolt to fly the modified Postal Monoplane to Cranwell. Macmillan saw them there checking inventories and signing clearances.

Now he noticed a great change: They looked 'hang-dog and dull-eyed.' According to Macmillan's own account, their eyes met and he saw in them the same expression as he had seen during the war on the faces of men he knew would die.

During that day, he passed and repassed them several times. Though they never spoke, there was now a bond between them. All three *knew*.

Macmillan left Northolt during the morning while they were still working. He and his wife had a lunch and tea engagement with a friend, and it was late that evening before they returned home.

'Squadron-Leader Jones-Williams and Flight Lieutenant Jenkins called at half past five,' their housekeeper told them. 'They asked to see the Captain.'

Neither had visited Macmillan's home before. He asked if they had left a message.

'They said "Just say we came to see Captain Macmillan to say goodbye." Then they said, "Good evening," and turned and walked slowly away. They looked to me, as you might say, like two ghosts.'

A red dawn was breaking on 17 December 1929 as Squadron-Leader Jones-Williams and Flight-Lieutenant Jenkins climbed on board the 'huge grey bird', as *The Times* described the Postal Monoplane. In the west, a full moon was glinting over the misty countryside. The plane had already been loaded with 1157 gallons of petrol, a sporting rifle, fifty rounds of ammunition, a revolver, enough iron rations to maintain life for three weeks, coffee in vacuum flasks, fruit and packets of sandwiches.

All up weight was 17,000 lb, and there was doubt at first whether the wind was strong enough for the plane to get airborne. Jones-Williams and Jenkins walked up the runway and looked at a valley of mist coloured pink by the early morning haze. They decided to risk an attempt, but would abort if they had not started to rise by the time they were level with the hangars.

In fact, they left the ground easily and smoothly, dipped into the valley to gather speed and then disappeared in the mist, leaving behind them for a few seconds the thrum of their Napier Lion engine.

The route weather forecast was ideal. A following wind would help them on their 4974-mile trip beyond the Cunene River in South-West Africa to beat the Frenchmen's long-distance record.

At midday they reported south of Paris, passed Marseilles at 2.27 pm and by four o'clock were fifty miles off the north-west coast of Sardinia.

A message should have been sent at 8 pm on 33.71 metres. None came. At midnight another message should have been transmitted but nothing was heard.

At lunch in the Fairey Aviation boardroom next day, they were discussing the Postal Monoplane's trip. Fairey always sat at the far end of the long dining-room, Macmillan at the other.

The test pilot still felt himself isolated. That lunch he talked little. There had been no further news of the aircraft, but radio conditions were bad. Platitudes were being exchanged: 'No news is good news. They're probably too far south for radio now.'

While the coffee was being served, Fairey's secretary came in and handed him a note. This was not unusual and nobody took any notice.

Then Fairey began to read out the note. 'Regret to inform you – ' He gave Macmillan a long look. 'Fairey Long-range Monoplane wrecked in mountains north of Tunis.'

Natives had led a French military patrol to the village of Zaghouan in Tunisia amongst mountains rising to 5000 feet, where they found the aeroplane and the bodies of the two airmen.

It was this point that they had to round before setting course for the Cape. They were doing exceptionally well, having achieved an average speed of 112 where only 86 mph was needed to break the record. As *The Times* said, 'only the direst ill-fortune could have brought such an accident. Everything so far had favoured the venture.'

That the shock of that accident had a lasting effect on Macmillan can be seen in his dedication of *Wings of Fate* thirty years later 'To four brave men I never forgot'. Two are his late brother and a pilot lost over the Atlantic with Princess Lowenstein in 1927 called Lieutenant-Colonel Minchin. The other two are 'the late Squadron Leader A.G. Jones-Williams, MC and Bar, Croix de Guerre, Royal Air Force' and 'the late Flight Lieutenant N.H. Jenkins, OBE, DFC, DSM, Royal Air Force'.

It seems that the bigger the disaster, the more 'shock waves' are felt before it happens. Before the disaster at Aberfan when the coal tip fell on the school and

144 people were killed, most of them children, there were thirty-five precognitive experiences. One of them was a nine-year-old girl who told her mother the day before the accident, 'I dreamed I went to school and there was no school there. Something black had come down all over it.'

There have been numerous premonitions of aircraft disasters. In 1967 a man phoned a London psychiatrist called Barker, who was researching premonitions, to say that he had had a vision of a large aircraft crashing into the side of an island, near a church surrounded by statues. He added that he was sure the accident was Nicosia in Cyprus and that 124 people would die.

The following month, a Britannia aircraft carrying service families tried to get in to Nicosia airport. There was low cloud and rain, with poor forward visibility. The pilot missed his approach first time, opened up the engines and went round again. The second time, he was closer but reported sudden low stratus. On his third approach, he tried to get in under the cloud and crashed, immediately killing himself and 123 others. Two other passengers died in hospital.

A Dutch psychic called Tholen suddenly saw the face of Queen Juliana floating over an aircraft behind which were forty-one funeral cars behind which one person walked. A week later the KLM constellation *Queen Juliana* crashed at Frankfurt. Forty-one people died, the stewardess being the only survivor.

Foretelling of the date of death, such as to Sefton Brancker,[1] is frequent, especially in eastern countries. On 5 April 1950 an RAF squadron-leader called Stiles was waylaid by a Sikh to have his future told for ten Singapore dollars. As proof of accuracy, the man persuaded Stiles to ask three personal questions. Being quite sure they could not be answered, he asked for the date of his birth, the date he joined the RAF and the date he was commissioned.

As all three answers were correct, Stiles had no option but to pass over the ten dollars. The Sikh said he could tell him two things that would occur in his future, but said he would prefer to withhold the third. Stiles insisted on his full money's worth and was told firstly that he and his wife would have a red-haired child, secondly that he would soon travel to Australia on business and thirdly, with reluctance but as he had insisted, that he would die of a heart attack in 1967.

Stiles discussed the forecast with his wife. The first was most improbable as they already had two children and had decided that was sufficient. The other two they had no control over, and they particularly worried about the third.

In March 1953 a boy was born to them with brown hair flecked with red. In 1957 he met a friend, a senior officer in the RAF, who told him that he had been selected for a special job on an island off Australia – but the posting had been cancelled because it was found he was now unfit to serve in the tropics.

That had been close, but the Sikh had been wrong. He might, Stiles reasoned, be wrong about 1967. On 24 May he had flown to Prestwick and back, interviewing candidates for permanent commissions. Not a particularly hard day, but that night he had a massive heart attack. When he came round in hospital, he was told by the doctor that he had had three separate terminals,

1 See 'Beyond the Inquiry'.

one of three minutes. He continued to live for thirteen more years, believing that technically, at least, the Sikh had been right.

Dr Barker, a psychiatrist who was so interested in precognition, wrote a book called *Scared to Death* which was published in 1968, the year he died. A year before, the man who foresaw the Nicosia crash had told Barker that he had a strong precognition of his death and urged him to take care and drive with caution. It had been unwelcome news, but like the true researcher he was, Barker kept up in a diary his own reactions to the prediction on the basis that it might cause interest and stimulate others to research further, if it did come true.

Air Vice-Marshal Sir Victor Goddard,[1] who had a distinguished career in the Far East during World War II, actually overheard a fellow officer speak about his death. At a cocktail party given in his honour in Shanghai in January 1946 he heard a naval officer say to another Englishman that they would not meet the guest of honour because he had died in an aeroplane crash the night before. Goddard went over and introduced himself, saying, 'I may be a bit moribund, but I'm not quite dead yet. What made you think I was?'

The commander then told him that he had had a dream in which an ordinary transport plane, 'Might have been a Dakota,' crashed in a snowstorm onto a rocky shore. On board were three English civilians, one of whom was a woman.

The plane that was to take Goddard to Tokyo next day was indeed a Dakota, but he was reassured because there was no possibility whatever of him picking up three civilian passengers, particularly the woman. Yet in the course of the dinner that followed, Sir Victor agreed to take Ken Seymour Berry of the *Daily Telegraph*. Later his host asked for a lift, which could hardly be denied. And finally, on behalf of the British Consul in Tokyo, he was cornered into also agreeing to take along a stenographer, Miss Dorita Breakspear.

The Dakota took off next morning. Flying over the mountains of Japan in a heavy snowstorm, the captain was forced down on to the shingle of a rocky Japanese island. Fortunately all survived – but it was an uncanny experience for Sir Victor.

Earlier in the 1930s he had had a strange precognitive experience of his own. In an article called 'Breaking the Time Barrier',[2] he describes a flight he made in an open-cockpitted aircraft with no radio or cloud-flying instruments. In heavy cloud he stalled and went into a spin which he was unable to correct. Losing height fast and aware that there were mountains around, he struggled with the normal recovery drill – stick fully forward and opposite rudder – emerging at two hundred feet into 'a murky sort of daylight', still spinning.

Suddenly he recognized the Firth of Forth, and regained his orientation, 'Thanks to the railings on the esplanade, and a girl who was running in pouring rain with a pram – she had to duck her head to miss my wingtips!'

Flying under the low cloud, he made for Drem airfield which he had visited the day before. After identifying the Edinburgh road, he saw looming ahead the black silhouettes of hangars. He crossed over the aerodrome boundary in deluging rain and dark and turbulent flying conditions.

1 Goddard was the original man selected to captain R101.
2 In *Light*, the journal of the College of Psychic Studies, Summer, 1966.

The next moment,

the airfield and all my immediate surroundings were miraculously bathed in full sunlight, as it seemed to me, the rain had ceased, the hangars were nearby, their north-west doors opened. Lined out in spick and span order on a newly laid tarmac were four aeroplanes; three bi-planes of a standard flying training type of aircraft called Avro 504N, one monoplane of an unknown type. We had at that time no monoplanes in the RAF, but the one I saw then was of the type which thereafter I carried in my memory and identified with the Magister which became, later on, a 'trainer'. Another peculiarity about the aeroplanes on the tarmac was that they were painted bright chrome yellow. All aircraft in the RAF in 1935 were exclusively aluminium-doped, there were no yellow aeroplanes. Later, because of an alarming increase in fatal accidents at flying training schools during the first phase of the expansion of the Air Force, the need became apparent for making training aircraft easily seen: in 1938 and 1939, more probably the latter, yellow aeroplanes became universal at all RAF flying training schools.

In the mouth of the hangar closest to me, another monoplane was being wheeled out. The mechanics pushing it were wearing blue overalls. As I passed over them, having climbed from only a few feet above ground to just high enough to clear the roof of the hangar, I must have been making a great deal of noise and, normally this would have caused a considerable sensation. Zooming the hangars, as I was doing, was a court-martial offence! It was quite certain that those mechanics must have looked up at me (had I been 'there' to them) as I flew over so close. But none of them looked up. This struck me as strange. It also struck me as strange that the airmen were wearing blue overalls. RAF mechanics had never worn anything else but brown overalls when working in hangars on aircraft. The hangar roofs, above which I was flying a moment later, were gleaming with the wetness of recent rain, but the bituminous fabric was entirely new and in very good order.

In 1939 Drem airfield was rebuilt and opened up as an elementary flying training school, where trainee pilots were instructed *ab initio* and sent solo. Drem was equipped with yellow 504N biplanes and Magister monoplanes, similar to the ones that Goddard had seen. Also airmen were now given blue overalls. In some strange way, he had leapt four years in Time.

10

The Rest
of the Room

$$\diamond\!\!\!\diamond\!\!\!\diamond\!\!\!\diamond$$

Time – it is in the air that lie hidden the clues to Time, where we can catch up with Time and begin to understand Time.

Within Time lie such mysteries as fate, clairvoyance, precognition, luck, time warps, apparitions.

The air has always fascinated mankind. And not only scientists but philosophers, poets and churchmen have shown extraordinary prescience about future happenings in the air. Leonardo da Vinci is well-known. Only a little later, Francesca Lana Terzi writes of an engine lighter than air which will raise itself in the air, and 'buoy up and carry into the air men or any other weight . . .'. He warns that 'no city can be secure against attack since our Ship may at any time be placed directly over it'.

The Bishop of Chester wrote in 1648, 'The volant or flying Automata are such mechanical contrivances as have a self-motion, whereby they are carried aloft in the open air, like the flight of birds. Such was that wooden Dove made by Archytas a citizen of Tarentum and one of Plato's acquaintance.'

Further on, the Bishop explains, 'Those ancient Notions were thought to be contrived by the force of some included air.'

The air has inspired insights and visions far ahead of their time. Airmen, more than most people, seem to accept the mysteriousness of the air and the universe beyond, perhaps because, faced with the awesomeness of time and space, they realize there is so much we do not know. Perhaps it is because, flying far above the world, they leave behind the everyday trials of life, and see civilization shrunk to its Lilliputian perspective. The aviator, Sir Alan Cobham wrote, 'lives in a world a little different from that of the ordinary man in the street. Perhaps it is that the "firebird" which St Exupery believed in is in all waiting for release, finds its release in flying.'

Throughout the ages, wisdom has been sought at altitude. The Oracle lies high in the mountains at Delphi. 'I will lift up mine eyes unto the hills, From whence cometh my help,' sang the psalmist. It was on the mountain that Moses was given the Tablets, in the mountain that John Smith found the Book of Mormon. Jesus Christ renounced the temptation of the Devil on the top of a high mountain and on the pinnacle of the Temple.

'You can do more clear thinking in two hours there than you can do in two days in a crowded city,' wrote a First World War pilot about being aloft in an aeroplane.

There is a phenomenon called 'cut-off' which can be experienced flying at great heights. How much this is a physical phenomenon and how much psychological, or whether the combination of height and speed allows the pilot to explore further unknown dimensions, is not completely known. The experience has been described as a half-way house to death – the feeling of isolation from the world is so intense that the pilot feels no sense of belonging to the earth at all, as if he were already a ghost.

Ghosts haunt a number of RAF stations. There were various reports of phantom bombers and fighters in World War II. North Weald, a fighter airfield in the Battle of Britain, had its own haunted hut – strangely enough number 101. But the most celebrated RAF ghost was at Montrose, on the north-east coast of Scotland.

The station was formed by the Royal Air Naval Service in 1918 to co-operate with the North Sea Fleet. Number 2 Squadron was there, but it was also a flying school. The instructors lived in a substantial manor house, the pupils – or 'Huns', as they were called, because they were reported to be better than the Germans at killing instructors – were housed in huts.

There were several versions of the story. According to Lord Balfour, who was Secretary of State for Air during the Second World War, a pupil at Montrose was young and nervous. His instructor considered it was high time he went on his first solo in a Farman biplane. But the pupil felt too frightened. The instructor insisted, and sent the boy off. At 300 feet, the boy tried to turn, stalled and dived into the ground. He was dead when he was pulled from the wreckage.

Balfour goes on, 'The instructor was asleep in his room that night when at 2 am he was woken by the feel of a presence. Beside his bed stood the figure of the dead pupil in his blood-stained leather coat. Not a word did the presence speak. He stood there in mute protest, an expression not of hate but of anguish on his face. The instructor was appalled. He let out a scream. Slowly the figure disappeared as it reached the door. Next day the instructor asked not only to move his room but to be posted away forthwith.'

A new instructor moved into the room. Three nights later the ghost appeared to him. He vacated the room hastily and a third man moved in. When the ghost appeared to him too, the commanding officer ordered the room to be locked and sealed.

According to C.G. Grey, the celebrated editor of the *Aeroplane*, the Montrose ghost was a fighter pilot who died in a crash. The aeroplane broke up in mid-air, and he was blamed for putting too much stress on it.

Some weeks after the inquiry which reached that verdict, an officer was going to the old mess when he saw a pilot in full flying kit walking ahead of him. This pilot reached the door, but did not open it. Nevertheless when he arrived, the officer found no trace of the pilot.

He saw the man, still in full flying kit, four or five times afterwards. The pilot still said nothing. Another officer in the middle of the night suddenly felt there was someone in his room. He raised himself and looked towards the foot of the bed. By the light of the dying fire, he saw a man in uniform sitting in a chair in front of the fire.

He asked him wno he was and what he wanted. There was no reply. He got out of bed and walked across the room. When he reached the chair, it was empty.

Several other officers saw the man, and two recognized him. He was identified as the pilot who had been involved in the aircraft cracking up. At that time unseasoned wood was being used in the construction and repair of fighters – and a second inquiry was instituted to examine this possibility as the cause of the crash.

A wing spar was found to have fractured because of a botched repair job on a joint, and the pilot error verdict was quashed. According to Grey, the ghost was never seen again. But after the Second World War, Peter Masefield nephew of the Poet Laureate, ex-chief executive of British European Airways and ex-managing director of Bristol Aircraft, reported giving a lift at Dalcross to a man in ancient kit. Masefield was flying south to an airfield in a light two-seater aeroplane, and the man got in the rear cockpit. When Masefield made a landing at Montrose, there was no sign of his passenger.

Masefield also reported coming in to land at a small airfield behind an old biplane. While he was on the approach, he saw the biplane stall and crash. Landing himself, he reported the accident and hurried over to where he was sure it had taken place. There was no sign of any aeroplane. Later he learned that some years previously a biplane had nosed in where he had seen the crash.

But the most famous of all apparitions in the sky are unidentified flying objects, known as flying saucers. Strange sights in the sky are often seen. They are often comets or shooting stars, the northern lights, the Aurora Borealis or other such phenomena. But there are some which do not appear to fit into any category we as yet know. Apparently inexplicable sights have been reported for centuries. They are the subject of legend, the origin, it is said, of the gods and the cult of the master race.

In biblical times, Elias was carried up to Heaven in a fiery chariot. Philip was flown to Babylon and back in a strange machine. In the sixth century a scribe wrote at length of a globe of fire flying over the countryside. The silver disc which appeared over Bylands Abbey in Yorkshire was seen by several monks and is documented.

The objects appear to come haphazardly through the centuries, though it is claimed they have vastly increased since the explosion of the first atom bomb, and since mankind has become apparently set on its planet's destruction. Towards the end of World War II, pilots reported being followed by what looked like glowing fireballs. These were christened foo fighters, and the phenomenon does not appear to be explained.

Ships at sea and aircraft at altitude have reported various strange disc-shaped objects which appeared to pass low over them or formate on them, and there have been thousands of sightings by people on the ground from many countries.

In 1947 the Air Technical Intelligence set up a Project Flying Saucer to examine the evidence. Many books have been written about them. In 1953 George Adamski wrote one book called *Flying Saucers have Landed*. He describes the UFO which he says he saw as a round metal disc with portholes in a circular deck-house above it.

In 1962 a flying saucer is reported as having landed on an aerodrome in Camba Punat in the extreme north-east of Argentina. And in 1969 many watchers in Europe saw two brilliant white lights beside the path of Apollo 12 on her way to the moon. One was in front and one behind, and they appeared to keep station with the spacecraft.

A big research programme called Project Blue Book was started at the University of Colorado, investigating and classifying reports, and in 1969 the lengthy Condon Report was issued. While it interpreted the vast majority of UFO sightings as imagination or misinterpretation of natural objects, it did not dismiss them all.

The Soviet Union is interested in UFOs, just as it is very seriously studying telepathy, precognition and extra-sensory perception. In 1967 Major General Stolyarov and a professor in astronomy appealed on television for any information of UFO sightings. They reputedly received thousands of letters in reply. Some were probably meteorites, is the favourite explanation.

Such is also the usual explanation of the Tungasky explosion of 30 June 1908. In the middle of Siberia, millions of trees were destroyed and marked by flashburns. The sound was heard 750 miles away. Aerial shock waves were recorded at all meteorological stations in the world. Next day there was such a strange bright light in the sky that people read newspapers at midnight. Animals were killed. But no one was hurt. Eyewitnesses reported, 'a huge black mushroom-like cloud to the north, covering the entire northern horizon', and the diameter of the crater was twenty-five miles. While the usual explanation is that it was a large meteorite, some Soviet scientists believe it might have been a nuclear device from outer space.

Several Soviet pilots have reported strange objects they met on patrol. Valentin Akkiratov saw a pearl-coloured circular object formating with him over Greenland. Believing it to be an American interceptor, he tried to shake it off. But the object kept station with him, finally climbing into the stratosphere and disappearing.

In 1961 a long deep scar was made along the shoreline of Lake Onega, near Povenets in the USSR. Witnesses reported a flying object low over the horizon which made no noise, though the impact with the ground was clearly heard.

On 20 April 1968 the crew of a Soviet Air Force plane reported,

a luminous object in the sky. It was red in colour and seemed to be flying at the same altitude, between five and six miles from their plane. From time to time, smaller luminous objects of the same colour appeared to approach the larger object, then wheeled around and darted away on a trajectory toward the Black Sea. In all, they observed four such satellites, each of which carried out the same manoeuvre. The display lasted five minutes, then the object disappeared.[1]

On 21 September 1968 a Soviet astronomer saw a brilliant crescent-shaped object in the sky that disappeared behind a mountain, which was also reported by many others. Silver disc-shaped objects that stopped car engines as they passed overhead have been observed both in America and the USSR. A farmer called Trent of McMinville, Oregon, took a photograph of a silvery circular

---

1 From *The New Soviet Psychic Discoveries*, Henry Gris and William Dick, Souvenir Press, 1978.

object about twenty-five feet in diameter, which he and his family observed and which the American Condon Report of Project Blue Book researched and described but did not explain.

On the night of 13–14 August 1956 a series of strange events were described and tracked at RAF stations Bentwaters and Sculthorpe and the big USAAF base at Lakenheath near Cambridge. The UFO was tracked by radar at all three stations. An RAF fighter took off and sighted round white rapidly moving objects which changed direction abruptly. The UFO appeared to circle behind the aircraft and followed it in spite of the fighter pilot's manoeuvrings. The Condon Report states, 'the preponderance of the evidence indicates the possibility of a genuine UFO in this case. The weather was generally clear with good visibility.'

People who have never seen a UFO naturally tend to treat with scepticism UFOs reported by people they do not know and have never heard of. It is different if you know the people concerned. In June 1954 the crew of a BOAC Stratocruiser saw a UFO on a flight from New York to London via Goose Bay, Labrador. I was a BOAC captain on the North Atlantic and knew most of the crew. Aircrew are not given to airy fancies and have seen too many things in the sky to be deluded into fantasy. Jim Howard was one of the quietest, steadiest, most reliable airline captains I ever knew. The idea of him producing a fairy-story is totally unrealistic, but when the very experienced Lee Boyd, the first officer, and the rest of that crew on the flight deck see what Jim Howard saw and confirm everything he reported, the evidence of the UFO's authenticity is overwhelming.

In Captain Jim Howard's own words, this is what happened. He writes,

I was in command of Stratocruiser GALSC on BOAC flight 510 from New York, Idlewild to London on 29 June 1954. My crew were First Officer Lee Boyd, Chief Engineer Officer Dan Godfrey, 2nd Engineer Officer Bill Stewart. The stewardess was Daphne Webster.

Lee Boyd was a distinguished Canadian Pathfinder Force Master Bomber, much decorated during WW2. Dan Godfrey was a veteran engineer from Imperial Airways days.

We departed Idlewild at 21.03 GMT (3 mins past 5 pm New York time), the daily 'Monarch' luxury flight to London with 28 passengers and 9 crew. I had planned a refuelling stop at Goose Bay in Labrador, after a flight of approximately 5 hours from New York.

Not long after take off from Idlewild, over Long Island sound, New York Air Traffic Control cleared us from their area and requested that we contact Boston ATC.

This we did and were instructed to go no further, but instead to hold at a specified point south-west of Boston, no reason being given. After entering the holding pattern my co-pilot and I discussed this delaying tactic, common enough when approaching a terminal area, but never before experienced by either of us departing such an area. I asked Boston to give the reason for the delay but was told to 'Standby'. After a further 10 minutes or so I pointed out to Boston that my fuel calculations had not allowed for such a delay so early in the flight and requested onward clearance. They then told me I could proceed if I would accept a diversion to the east of Boston, over Cape Cod, to rejoin our planned route some 100 miles or so northeast of Boston – no reason being given. I accepted and was told to proceed on my way.

The flight then went smoothly, mostly on top of broken cloud, altitude 18,000 feet, airspeed 230 knots.

Crossing the estuary of the St Lawrence River near Seven Islands, Quebec, I noticed down on the left side of the aircraft, visible through gaps in the cloud, some objects moving along parallel to our track, about 3 miles or so from us. At first I thought they looked like balloons but obviously they could not have been since they seemed to be keeping pace with our aircraft. I turned to Lee Boyd in the co-pilot's seat, to point them out, but found that he had seen them too. As we crossed the north bank of the river, into Labrador, the cloud began to thin out and soon disappeared completely leaving these things in full view.

The sun was low in the northwest but still about half an hour before sunset, visibility excellent. As we watched (the rest of the flight-deck crew had by now spotted them too) they climbed steadily up until they appeared to be at the same altitude as ourselves, maintaining station off our port wing, distance unknown but possibly a mile or two from us. There were seven objects in all, one large, six small, strung out in a horizontal line. The small ones moved steadily in relation to the larger, so that sometimes there were 2 ahead, 4 behind, sometimes 3 ahead, 3 behind, and so on. The large object changed its shape several times, gradually assuming different silhouettes. The smaller ones always looked 'globular'. The colour appeared to be dark grey, and silhouetted as they were against a bright sky, sharp-edged and opaque. I sketched the appearance of them several times during the 20 minutes that they flew with us, covering some 80 miles during this time. I did not alter course, the aircraft was on auto-pilot throughout, heading for Goose Bay.

After a little while, the stewardess came on to the Flight Deck and told me that the passengers were all watching these things and could I tell them what they were! I couldn't.

I then asked Lee Boyd to contact Goose Bay radio and tell them what was happening. They responded by saying that they had a radar equipped night fighter flying in their vicinity and suggested he should head in our direction to have a look. I agreed to this and after a few more minutes changed radio frequency in order to talk directly to the fighter pilot. As he spoke, saying he had us in radar contact about 20 miles ahead, closing fast, the small objects seemed to close into the large one and disappear. A few moments later the fighter pilot said he would soon be above us, 'What did they look like now?'

I told him there was now only one left and gave him a bearing to steer. As I did this the object seemed to shrink quite rapidly until it had completely gone.

By now we had already passed our normal time for start of descent into Goose Bay so I changed frequency back to the Control Tower and started down.

We landed at 01.51 GMT just after sunset, where we were questioned by a United States Air Force officer – who did not seem in the least surprised at what we had to say – indeed he implied that such sightings were commonplace in that area!

The rest of the flight to London was uneventful, arrival time 1.30 pm on 30th June.

These are the facts of the sighting as they happened – later I learned several things that had a bearing on the matter but did nothing to explain it.

As a result of an unknown leak to the press of my voyage report to BOAC headquarters, the sighting was given a great deal of publicity around the world, and I received a great deal of mail. One letter was from an American doctor who, at the time of our sighting was on holiday with his wife in a lakeside cabin in Massachusetts. Early in the evening of 29th June, they both heard a strange roaring sound and saw flying over the lake in a North Easterly direction one large and six small objects. He and his wife (he claimed) immediately drew sketches of what they had seen which showed remarkable similarity to my sketches.

Question. Did Boston Air Traffic Control have these strange things on their radar screens, and was this why we were held? I shall probably never know, for when the BBC researching for a TV documentary on UFOs asked Boston to provide information, they were told that the records had been destroyed. Odd? Maybe – maybe not; several years had elapsed between the sighting and the BBC enquiry.

Again, some years later I was shown a copy of a reported sighting of 7 UFOs made by a Canadian Govt. Survey Vessel in Hudson Bay on the 29th June 1954. The appearance of the objects was similar to the things we saw.

One question people ask when they hear this tale is, 'How did you feel? Weren't you scared?'

I talked this over with other members of that crew later, and we all agreed that very far from being scared, we all felt a powerful feeling of attraction towards these things and experienced a sense of loss, of desolation, when they disappeared. It perhaps sounds unreal, but the feeling was so strong that it has stayed vividly in my memory for the 27 years since it happened.

Perhaps this is the special gift of an airman. Perhaps an airman's mastery of a third dimension allows him brief glimpses of others much vaster and as yet unconquered. Perhaps there is an evolving pattern. Mysteries within mysteries which are solved only to expose much more profound ones.

The mystery of the R101 which first brought the medium Eileen Garrett into worldwide prominence still discloses other dimensions. Hugh Dowding, who agreed the R101's airworthiness certificate, had only just been appointed Air Member for Supply and Research, the decision to go to India had already been taken before his appointment and, new to airships, he was heavily dependent on the advice of experts. But he came to the conclusion in later years that he should have insisted on more tests. A man of great courage and integrity, he was not to make the same mistake again.

If any single person could be said to have turned the tide of World War II, it was Hugh Dowding. It was *his* strategy, his courageous opposition to Winston Churchill which caused the fighter squadrons he commanded to win the vital Battle of Britain, a victory which historians have compared with the Battle of Trafalgar.

Yet the rewards of Nelson did not come Dowding's way. He was sacked from his position soon after the battle. He did not receive public acclaim. But what concerned him more than acclaim was the sorrows he witnessed in the families of wartime pilots. This led him to investigate spiritual matters. Like everything he did his investigation was thorough. He became convinced.

In his foreword to Basil Collier's biography, *Leader of the Few*, Dowding writes, 'The evidence for the conscious survival of death, and the possibility of intelligent communication between the quick and the dead is in my opinion quite convincing to the open mind.'

Lord Dowding was also convinced of the guiding hand of Destiny. That feeling of destiny would be difficult to dismiss in the story of his marriage to Muriel, the present Lady Dowding. When one of Lord Dowding's closest friends asked him, some years after the war, if he felt no bitterness about his treatment by the government, adding that even God had not rewarded him, Lord Dowding replied, 'But God did. He gave me Muriel.'

Muriel was, and still is, a very beautiful, vital woman. She had been brought up in comfortable circumstances, a gay, apparently spoiled flapper in the early 1930s. However, spoiled she was not.

She too was convinced of the hand of Destiny. As a little girl she had suffered from nightmares. Sometimes her father comforted her. But at the age of ten she was aware of a man who came to her within the dream itself. He was tall and slim with blue eyes and hair just beginning to grey at the temples. He was a soldier dressed in khaki. She knew his name was Hugh and that she would marry him.

Yet in 1934 she met a rich dashing young man called James Max Whiting who fell deeply in love with her. He refused to take no for an answer. They were married in August 1935. Their son was born in 1938, just as it was becoming clear that war was inevitable. Max Whiting was an engineer and had studied in a German electronics factory. Therefore when war came he was in a reserved occupation. Nevertheless he secured his release and joined the RAF. He served as an engineer on a Lancaster squadron, their targets mainly deep in Germany. On 22 May 1944 Max Whiting was posted missing, and Muriel received the official telegram. It had just been given out over the BBC news that Lancasters of Bomber Command had raided Duisberg and Muriel assumed it was on this raid he had been shot down.

Like many airmen, Max Whiting had left a letter to be sent to his wife in case he did not return from a raid. It said, 'If you ever receive this, it means I have not been able to get back to you. But don't worry . . . I shall turn up. If I am able to do this, I will try and contact you through the sources in which you believe.'

The next month, Muriel had what she described as a strange half-waking dream. The tall soldier of her childhood dreams came into her bedroom. But he was no longer dressed as a soldier. He was dressed in grey flannel trousers, blue shirt and wore a black knitted tie.

As with Captain Howard and the UFOs, she was aware of a powerful feeling of attraction. She realized he was someone she knew and loved. He was standing by the mantelpiece arranging some ornaments which were black elephants, and later she discovered that black elephants are symbolic of happiness. Even to this present day she can remember that feeling of happiness, and of calling out his name, 'Hugh', of his turning and smiling at her.

The others who were staying in the cottage heard Muriel laughing and talking, and when they went into her room, she was wide awake, smiling as though she had come to terms with her sorrow.

A month later, she went to stay with her in-laws, and they suggested she consult a medium. Muriel went up to London, to the College of Psychic Studies (Sciences). Through Miss Topcott, the medium, Muriel was convinced she spoke to the spirit of her dead husband. He told her he had been pursued and shot down by German fighter pilots off the coast of Norway.

Muriel Whiting protested that he went to Duisberg, but the voice she believed to be Max's had replied, 'I will try to get you the evidence. But I assure you, the last I saw with my physical eyes was the coast of Norway.'

No evidence was forthcoming. The anguish of not knowing what had happened prompted Mr Whiting Senior to write to Lord Dowding, who was known to be active in getting what information he could for the families of airmen. Just before Mr Whiting sent the letter, he said to his daughter-in-law, 'It would come better from you.'

Therefore with only one or two minor alterations, she copied out the letter and posted it. She received a kind acknowledgement. Then nothing.

Some months later she was delighted to receive another letter from Lord Dowding saying he rather thought Max had made contact through a recent spiritualist sitting. He invited her to meet the medium, and for this purpose suggested she lunch with both of them at the United Services Club in Pall Mall.

When Muriel arrived all her thoughts were on possible contact with Max, and all her attention on the medium Mrs Hunt. She hardly noticed the tall grey-haired gentleman who was Lord Dowding. Mrs Hunt naturally did most of the talking over lunch, telling her about the sitting where Max had apparently come through, but afterwards Mrs Hunt had to leave for another appointment.

When she had gone Lord Dowding said, 'If you don't have to go at once, will you stay and have tea with me?'

As he came round to move back her chair, Muriel looked at him closely for the first time. She says in her book,

In that instantaneous glance, I knew. He was the person I had known as a little girl, the figure in khaki. . . .

In my astonishment at this sudden rush of recognition, I gasped, 'Hugh!'

Then before I could even begin to feel embarrassed at having been so over-familiar, he said quietly and kindly: 'How charming of you to call me that.'

Some years later after they were married, Muriel asked him if he had invited all the war widows who wrote to him out to lunch and he replied gravely, 'No. Only you. Because at that sitting, Max asked me to.'

And the evidence Max had promised when he spoke to her through the medium? That too came through from a totally unexpected source.

Late in 1946 Muriel's step-father, Mr Farrant, was travelling by train to Tunbridge Wells. The only other passenger in the compartment was a Dane. They began talking and naturally in those days the main subject of conversation was the war. Mr Farrant asked the Dane how he had fared in those awful years.

The Dane told him of one great tragedy. He explained, 'I own a farm that runs down to the sea. I shall not forget the night of 22 May 1944, when we heard one of your great Lancaster bombers fly over towards Norway. The German planes came after it and it was shot down. We were able to salvage part of the plane and most of the bodies and we built a little memorial to them.'

The Dane then brought out of his pocket a photograph of the memorial. It was of a single large upright stone, rough hewn, which was surrounded by smaller rounded stones in a simple cairn. The upright was inscribed in Danish, 'To our gallant English Allies.' Around it on the faces of the individual stones were carved the names of the crew. Among them clear and distinct was J.M. Whiting, Kent.

The photograph is now in Lady Dowding's possession. The mystery of what the lone Lancaster was doing has not yet been solved. But it was off the coast of Norway, as Max Whiting said.

Lord Dowding, concluding his preface to *Leader of the Few*, writes of communication with the dead, flying saucers and other phenomena, 'I confidently predict that all these ideas will be commonly accepted in a hundred years time, when those who reject them will be classed with those who now believe that the earth is flat.'

It is of those airmen like the Wright brothers, Antoine St Exupéry, Sir Sefton Brancker, Lord Dowding, Captain Foote, Sir Victor Goddard, Captain Howard, Colonel S.F. Cody, J.W. Dunne and many others, of scientists like Cayley, Chanute and Lilienthal, of seers like Leonardo da Vinci, Francesca Terzi and the Bishop of Chester that the poet Vine Hall writes:

> Therefore, I see,
> As knit in one society,
> Seers, saints, and airmen, all who rise
> To the pure place of the clear skies
> And read as in a mirror's face,
> The hidden things of time and space.

The view through the keyhole is still limited – but every now and again come tiny chinks of light. And if we have examined new dimensions or admitted the possibility of new concepts, we have made the keyhole a little wider and allowed more light to come in.

Bibliography

Abbott, Patrick. *Airships*. Adams and Dart, 1973.
Babington-Smith, Constance. *Amy Johnson*. Collins, 1967.
Balfour, Harold. *Wings Over Westminster*. Hutchinson, 1973.
Barker, Ralph. *Great Mysteries of the Air*. Chatto and Windus, 1966.
—. *Verdict on a Lost Flyer*. Harrap, 1969.
Beaty, David. *The Human Factor in Aircraft Accidents*. Secker and Warburg, 1969.
—. *The Water Jump*. Secker and Warburg, 1976.
Bellis, Hannah. *Amy Johnson*. Newnes, 1953.
Burke, John. *Winged Legend*. Barker, 1970.
Bryden, H.G. (ed.). *Wings: Anthology*. Faber and Faber, 1942.
Carter, I.R. *Southern Cloud*. Angus and Robertson, 1964.
Churchill, Winston. *The Second World War*. Cassell, 1950.
Collier, Basil. *The Airship*. Granada, 1974.
Daley, Robert. *An American Saga*. Random House, 1980.
Davis, Pedr. *Charles Kingsford Smith*. Summit Books.
Dean, Maurice. *The RAF in Two World Wars*. Cassell, 1979.
Dixon, Norman. *Preconscious Processing*. John Wiley, 1981.
Dowding, Muriel, Lady. *Beauty Not the Beast*. Spearman, 1981.
Earhart, Amelia. *20 Hrs 40 Min*. Putnam, 1928.
Ebon, Martin. *Prophesy in Our Time*. Signet.
Eckener, Hugo. *My Zeppelins*. Putnam, 1949.
Ellison, Norman. *Flying Matilda*. Angus and Robertson, 1957.
Elwood, Gracia Fay. *Psychic Visits to the Past*. Signet, 1971.
Fish, Donald. *Airline Detective*. Collins, 1962.
Fitzgibbon, Louis. *Katyn*. Stacey, 1971.
Foote, Wilder (ed.). *The Servant of Peace: Dag Hammarskjöld*. Bodley Head, 1972.
Fuller, John. *The Airmen Who Would Not Die*. Putnam, 1979.
Goddard, Victor. *Flight Towards Reality*. Turnstone Press, 1975.
Goerner, Fred. *The Search for Amelia Earhart*. Bodley Head, 1966.
Greenhouse, Herbert B. *Premonitions*. Turnstone, 1971.
Grey, Elizabeth. *Winged Victory*. Houghton Mifflin, 1966.
Gris, Henry and William Dick. *The New Soviet Psychic Discoveries*. Souvenir Press, 1978.
Hansard, Lords' Debate, 18 October 1961.
Haslam, E.B. *RAF Quarterly*, Vol. 12, Nos 2 and 3.
HMSO, *Merchant Airmen*.

Hoehling, A.A. *Who Destroyed the Hindenburg?* Robert Hale, 1962.

Irving, David. *Accident*. Kimber, 1967.

Jahoda, Gustav. *The Psychology of Superstition*. Penguin, 1969.

King, Alison. 'Amy's Last Flight' in *The Aeroplane*.

Knight, R.W. (ed.). *The Hindenburg Accident: Official Summary of Enquiry*. United States Department of Commerce, Safety Department, 1938.

Koestler, Arthur. *The Case of the Midwife Toad*. Hutchinson, 1971.

—. *The Roots of Coincidence*. Hutchinson, 1972.

—. Arnold Toynbee, et al. *Life After Death*. Weidenfeld and Nicolson, 1971.

—. Alistair Hardy, Robert Harvie. *The Challenge of Chance*. Hutchinson, 1972.

Lash, Joseph. *Dag Hammarskjöld*. Cassell, 1962.

LePoer, Brinsley. *Chariots of Yesterday, UFOs of Today*. Souvenir Press.

Macmillan, Norman. *Wings of Fate*. G. Bell, 1967.

McCann, Lee. *Nostradamus*. Creative Press, 1941.

McKee, Alexander. *Into the Blue*. Souvenir Press, 1981.

McNally, Ward. *'Smithy'*. Robert Hale, 1966.

Meager, George. *My Airship Flights*. Kimber, 1970.

Milburn, Irene. *Runways to Adventure*. Robert Hale, 1960.

Mollison, James. *Death Cometh Soon or Late*. Hutchinson, 1932.

—. *Playboy of the Air*. Michael Joseph, 1937.

Mooney, Michael. *Hindenburg*. Hart-Davis, McGibbon, 1972.

O'Brien, Conor Cruise. *To Katanga and Back*.

Putnam, George (ed.). *Amelia Earhart: Last Flight*. Putnam, 1937.

Ramsden, J.M. *The Safe Airline*. Macdonald and Jane's, 1976.

Smith, Elinor. *Aviatrix*. Harcourt Brace Jovanovich, 1981.

Taylor, P.G. *Pacific Flight*. Angus and Robertson.

Titler, Dale. *Wings of Mystery*. Dodd Mead.

Toland, John. *Ships of the Sky*. Holt, 1957.

Twigg, Ena. *Medium*. Hawthorne, 1972.

Urquhart, Brian. *Hammarskjöld*. Bodley Head, 1972.

Warsaw in Chains. Allen and Unwin, 1959.

Williams, Captain. *Airship Pilot No. 28*. Kimber, 1974.

Woodworth, Robert and Harold Schlosberg. *Experimental Psychology*. Methuen, 1967.

Woolf, Virginia. *Diary*, Vol. III (1925–30). Hogarth Press.

Zichy, Count. 'There But for the Grace of God' in *Aeroplane Monthly*, March 1981.

Also the *Herald* (Melbourne), *Sydney Morning Herald*, *The Times*, *Daily Mail*, *Daily Express*, *Flight*, *Daily Telegraph*, *Sunday Despatch*, *Bedford Times*, *The Aeroplane*, *Jane's*, *Hansard*, ICAO Accident Reports, HMSO and US Accident Reports and *Air Mail*.

Index

Aberfan, 138–9
Adamski, George, 144; *Flying Saucers Have Landed*, 144
Aeroplane magazine, 110, 143
Air Council, 22
Air League, 72
airmail service, 87, 89
Air Ministry, 11, 27, 30, 33, 35, 39–40, 42–3, 52, 60, 136
airships:
   Hindenberg (LZ129), 45–51
   R34, 24
   R38, 49
   R100, 21, 23–4, 29, 41, 43
   R101, 13, 15, 20–46, 50–1, 101, 148
   R102, 43
   Skyship 500, 44
   Zeppelin, 21, 45–51
Airspeed Envoy, 89
Airspeed Oxford, 70–2, 74–8; Oxford V.3540, 70–2, 74–8
Air Technical Intelligence, 144
Air Transport Auxiliary, 69, 74
Akkiratov, Valentin, 145
Albatross, HMS (aircraft carrier), 80
Alcock, Sir John, 24
Allan, Scotty, 80, 83–4, 91
All-Red-Route, 24
Altair, see Lockheed
Ambrosi, A., 48
America, see Friendship
American Nurse, The, 20
Ames (psychologist), 119
Amulree, Lord, 42

Anderson (friend of Kingsford Smith), 79–80, 87, 94
Anson, see Avro
Apollo XI, 52; Apollo XII, 145; Apollo XIII, 52; Apollo XIV, 16
Armstrong Whitworth Argosy, 33
Argosy, 33
astrology, 13–14, 19
Atcherley, R.L.R., 72
Atherstone, Lt. Cmndr., 25, 28, 32, 35
Atlantic Ocean, 20, 23–4, 45, 47, 49–50, 53, 56, 60, 63, 81, 87, 98–100, 108–9, 129, 138
Australian National Airways (ANA), 60, 80–3, 85–6, 94
Automobile Club, 81
Avro aircraft:
   504N, 141
   Anson, 74
   Avian, 87–8, 92
   Lancaster, 104, 149–50
   Shackleton, 115
   Tudor, 12
   Type X, 92
   York, 11–14, 17–20, 108, 116–17; G-AHFA, 11–14, 17–20

BALPA (British Airline Pilots' Association), 112–16
BALPA Air Safety and Technical Committee, 114
Bailey, Lady, 72
Bairstow, Prof., 27
Balfour, Lord, 52–3, 143.

Balfour Boeings, *see* Boeing 314
Barker, Dr, 138–40; *Scared to Death*, 140
Bard, Monsieur and Madame, 33
Battle of Britain, 43, 52, 148
BEA (British European Airways), 12, 144
Beardmore Tornado diesel engine, 25, 34
Beauvais, 33, 35–6, 38, 44
Beaverbrook, Max, 52–3
Bedford, 20–1, 35, 39, 45
Bell, Joe, 25, 29–30, 32, 34, 44, 101
Berkeley, HMS, 75, 77
Bermuda Triangle, 18, 68
Berwick, 53–4
Bibesco, Princess Marthe, 30, 35, 43–4
Binks, Arthur, 25, 29, 32, 34, 101
bio-rhythms, 14
Bishop (Air Ministry inspector), 30
Bishop of Chester, 142, 151
Black Magic, 74
Blériot, Louis, 133
BOAC (British Overseas Airways Corporation), 54–5, 95, 98, 102, 104–19, 146
Bodenan, Francis, 128
Bodensee, 46
Boeing Co., 112
Boeing aircraft:
  314 flying boat, 52–6
  Stratocruiser, 95–102, 146; 90944, 98; 90947, 97; G-ALSC, 146–8; G-AKGM (Golf Mike), 98–100, 102; G-ANTY, 102; N90943, 97–8
Bolton, Mrs Frances, 44
Bomber Command, 149
Booth, Capt. Ralph, 23, 28–9, 41
Boyd, Lee, 146–7
Brabazon, Lord, *see* Moore-Brabazon
Brancker, Sir Sefton, 21, 26, 28, 33, 35, 44, 135, 139, 151
Brancker, Lady, 42
Brazilian Boundary Commission, 96
Breakspear, Dorita, 140
Bridge Hotel, Bedford, 24, 29, 39
Bristol Aircraft Co., 144
Bristol aircraft:
  Brabazon, 43
  Britannia, 115, 117, 139
  VC10, 118
Broadbent (airman), 89–90
Brown, Sir Arthur Whitten, 24

Brown, Rosemary, 1
Buck (Lord Thomson's valet), 30–1
Bunker, Betsy, 24, 29
Bushfield (Air Ministry inspector), 30
Byatt, Robert, 85
Byrd, Comdr., 60

Calcutta, 110–12, 114–15, 117
Canadian Pacific Airlines, 108–9
Capper, Col., 133
Cardington Royal Airship Works, 21–7, 29–30, 32, 35, 38, 40, 43–5
Cayley, Sir George, 133, 151
Cazalet, Victor, 131
Cellon, 22
Chanute, Octave, 151
Charlton, William, 143
Chopping (navigating officer), 11
Church (rigger), 29, 34, 36, 42
Churchill, Sir Winston, 15, 17, 52, 54, 62, 121, 148; *The Second World War*, 54–5
Ciampino Airport, Rome, 104–5, 113, 118–19
Clarke, Haden, 88
Cobham, Sir Alan, 142
Cody, Samuel Franklin, 73, 133
'collective unconscious', 17, 101–2
College of Psychic Studies, 149
Collier, Basil, 148; *Leader of the Few*, 148, 151
Colmore, Wing Comdr., 21, 23, 25–9, 35, 40
Colmore, Mrs, 42
Columbia, 52
Columbird, 52
Comet, *see* De Havilland
Commonwealth Conference, 21, 24, 26, 28
Conan Doyle, Sir Arthur, 25, 36, 62
Condon Report, 145–6
Congo, Republic of the, 121–9
Consolidated Liberator, 109, 130–1
Constellation, *see* Lockheed
Convair, *see* General Dynamics Corporation
Cook (R101 crewman), 34
Coster, Ian, 36–7
Cranwell RAF College, 61–2, 137
Cripps, Sir Stafford, 38
Croft, Capt. Val, 98–102

Cross (shepherd), 85, 92
Croydon, 73

Daily Telegraph, 55–6, 115, 119, 140
Dakota, *see* Douglas
Darling, George, 35
Davis, Pedr, 93
Da Vinci, Leonardo, 133, 142, 151
Dawn, The, 20
DC4/6, *see* Douglas
De Havilland Co., 104–5, 110, 116–18
De Havilland aircraft:
    Comet, 104–20; I, 104–8, 115, 117;
      IA, 117; II, 112, 116–17; III, 112,
      116–17; IV, 116–17, 119; Yoke
      Peter, 113–16; Yoke Uncle, 116–17;
      Yoke Victor, 110–15; Yoke Yoke,
      115; Yoke Zebra (G-ALYZ), 104–8
    Gipsy Moth, 73, 86
Denison House, Boston, 63–4
Detroiter, see Southern Cross
Disley (electrician), 29, 34
DNA, 15, 17
Dobbs, A., 101
Douglas aircraft:
    Dakota, 111, 140
    DC4 00-RIC, 123–4
    DC6B, SE-4BDY, 123–8
Dowding, Lord, 17, 27, 31, 43, 148–51
Dowding, Muriel Lady, 148–50
Drem airfield, 140–1
Duchess of York, 30
Duke of York, HMS, 53
Duke University, Parapsychology
    Department, 16, 44, 65
Dum Dum Airport, Calcutta, 111
Dunne, J.W., 133–4, 151
Dunnell, Charles, 81–2, 94
Dunnell, Mr (father of Charles), 82
Dyer (shepherd), 85, 92
Dymond, Harold, 91

Earhart, Amelia, 63–9, 88; *20 Hrs, 40
    Mins*, 64
Eckener, Capt. Hugo, 45–6
Eddington, Sir Arthur, 101
Einstein, Albert, 101–2, 134
Electra, *see* Lockheed
Elizabethan, 12
Empress of Hawaii, 55, 108–10, 113
Endeavour, 20, 61–3
Evening News, 72–3, 115

Ewart, Mrs, 92
extra-sensory perception, 16, 65
Eysenck, Hans, 14

Fairey Aviation Co., 135, 138
Fairey, Dick, 135–6, 138
Fairey aircraft:
    Fox, 135
    Postal Monoplane, 135–8
Faith in Australia, see Southern Moon
Farman biplane, 143
Farnborough Royal Aircraft
    Establishment (RAE), 112, 116, 133
Farrant, Mr (Lady Dowding's step-
    father), 150
Farrall, Hubert, 82, 92
Farrall, Mrs Hubert, 92
Feischner, Dr, 36
Fighter Command, 55
First British Airborne Division, 18
Fleming, Dr Archibald, 36
Fletcher, Comdr., 75, 77
Flight magazine, 38
flying boats, 49, 52–5, 65, 90
Fokker, 60–1, 63, 80–94
Foote, Capt. Harry (Robert), 104–5,
    107–8, 110–20, 151
Foote, Mrs Harry (Chris), 118–20
Fox, *see* Fairey
Frankfurt, 45, 49, 139
French Air Force, 36, 122
Friendship (formerly *America*), 60, 63
Friendship, SS, 45

Gander, 11–12, 18
Garrett, Eileen, 25–6, 35–6, 39–40,
    43–4, 62, 148
General Dynamics Corporation Convair,
    16–17, 38
Gent (engineer), 25
George V, 30
George V Jubilee airmail, 89
George VI, 46
German Air Ministry, 47, 49
Ghana Airways, 102
Ghost jet engine, 104
Giblett (meteorological officer), 29, 33
Gibraltar, 130–2
Gibson, Capt. Alan, 113–14
Gipsy Moth, *see* De Havilland
Glasgow, May, 82, 91

Goddard, Sir Victor, 15, 28, 44–5, 140–1, 151; 'Breaking the Time Barrier', 140
Godfrey, Dan, 146
Goebbels, Dr Joseph, 46
Goering, Hermann, 46, 49–50
Goerner, Fred, 67
Goldsborough (navigator), 61
Goodrich Co., 63
Goodrich Silvertown aircraft, 63, 91
Goose Bay, 146–7
Gordon (mechanic), 64
Graf Zeppelin, 45–6
Grayson, Mrs Frances, 61, 63–4
Gregory Air Service, 128
Grey, C.C., 143–4
Grey Walter, Dr, 14
Grossack, Capt., 95
Guest, Frederick, 60
Guest, Mrs Frederick, 60, 63–4
Gulbransen, Capt. Hank, 99–102

Haddon, Capt. Maurice, 111
Hall, Sir Arnold, 116
Hall, Vine, 151
Hallonquist, Capt., 123, 216
Hamilton (airman), 60–1
Hammarskjöld, Dag, 121–9
Hancock, Capt., 80
Handley Page Hermes, 56–9, 114
Haslemere, HMS, 74–6
Hardy, Thomas, 101
Hawker Hurricane, 55
Hawker Siddeley, 125, 128
Hayes, 136
Hayter, Mrs (daughter of Shortridge), 92
Heath, Lady, 64, 72
Heathrow Airport, 104, 109
Hendon Air Display, 23
Henlow, 28, 38, 71
Hermes, *see* Handley Page
Higgins, Air Vice-Marshal Sir John, 23, 27, 136–7
Hinchliffe, Capt., 25, 35–6, 61–3, 81
Hinchliffe, Emilie, 25, 35, 61–3
Hindenberg, *see* airships
Hinkler, Sqn. Ldr., 73, 95
Hitchcock (airman), 94
Hitler, Adolf, 45–7, 49, 129
Hochhuth, Rolf, 131–2; *Soldiers*, 132
Holden, Capt., 87, 94

Hollis, Gen. Sir Leslie, 53–4, 55
Hood, Clyde E. (Charles), 81, 92
Howard, Capt. Jim, 146–9, 151
Howden, 21
Howland Island, 66–8
Humphreys, Charles, 74
Hunt, 'Sky', 23, 27, 34–5
Hunt, Mrs (medium), 150
Hurricane, *see* Hawker

Idalia, 79
IFALPA (International Federation of Airline Pilots), 117–18
imagination, 52–9
Imperial Airways, 49, 53, 81, 87, 146
Inchcape, Lord, 61
Indian Civil Aviation Authority, 30, 111
Indian Express, 125
Inglis, Prof. C., 38, 42
Institute of Aeronautical and Mechanical Engineers, 81
Irwin, Flt. Lt. Herbert Carmichael, 21–3, 26, 28–9, 35–42, 45
Irwin, Olive, 23, 28, 35, 38, 41–2
Itasca, USS, 66

Jasmin, 73
Jenkins, Flt. Lt. N.H., 136–8
Jeppesen's manual, 126
Johnson, Amy, 64, 69–78, 88, 93
Johnson, Sqn. Ldr., 24–5, 29
Jones-Williams, Sqn. Ldr. A.G., 136–8
Jung, Carl, 17, 101

Kano, 57–9
Kammerer, Paul, 101
Karachi, 27, 32, 109–10, 113, 117–20
Kelly-Rogers, Capt. Jack, 53–6
Kidlington, 70–1, 76
King (engineer), 25, 29, 35
Kingsford Smith, Sir Charles, 60, 79–93
Kingsford Smith (brother of Charles), 90
KLM Constellation, 112, 139
Koestler, Arthur, 19, 101
Krebs, Herr, 45

Lady Southern Cross, 87, 79–91, 93–4
Lakehurst, NJ, 47–8
Lancaster, *see* Avro
Lancaster, Capt. Bill, 67, 88, 93–4
Lancaster, Mrs, 88
Lansdowne, Lord, 123–4

Leasor, James, 55; *War at the Top*, 55
Lee, Auriol, 26, 33
Lee, Lincoln, 14; *Three Dimensional Darkness*, 14
Leech, Harry, 32–4
Lehmann, Capt., 47–50
Lehmann, Frau, 47
Lend-Lease Act 1941, 44
Lexington, USS, 67
Liberator, *see* Consolidated Liberator
Lilienthal, Otto, 151
Lindbergh, Charles ('Lucky'), 20, 60, 63–6, 79
Littlejohn (airman), 90
Lloyd George, David, 60
Lockheed Co., 65
Lockheed aircraft:
   Altair, 87, 91, 93
   Constellation, 68, 95, 104, 109, 112, 115, 139
   Electra, 65–7
   Vega, 65
London Aeroplane Club, 72
London Airport, *see* Heathrow
Lowenstein-Wertheim, Princess, 60, 63, 138
Lumumba, Patrice, 121, 128
Luther, Dr, 46

MacDonald, James Ramsey, 21, 31, 35, 43
Macfarlane, Gen. Mason, 130
Mackay, Hon. Elsie, 25, 61–3
Macmillan, Capt. Norman, 135–8; *Wings of Fate*, 138
McWade, F., 22–3, 40–1
Magister, *see* Miles Magister
Maisky, M., 129–31
Majendie, Capt. Michael, 105–7, 114–15, 118
Manning, Capt., 65–6
Margules, Julian, 82
Martin flying boat, 65
Masefield, Peter, 144, 151
Mason (engineer), 42
Maugham, W. Somerset, 135; *Sheppey*, 135
May, Capt. Bill, 108–9
Maxaret brakes, 116
Mead (*Southern Cloud* witness), 85
Megginson, 'Carbolic', 34
Melrose (airman), 89–91, 93

Mew Gull, *see* Percival
MI5, 129
Middleton, Drew, 74
Milburn, Irene, 110
Miles Magister, 141
Miller, Mrs Chubbie, 67, 88, 93
Minchin, Lt. Col., 60, 138
Ministry of Civil Aviation, 108, 114–17
Miss Southern Cross, 87, 89, 94
Mollison, Jim, 73–4, 80, 84, 86–8, 93–4; *Death Cometh Soon or Late*, 93
Montreal, 23–4, 29, 60
'Montrose ghost', 143–4
Moore-Brabazon, Col. (later Lord), 38, 42
Moran, Lord, 55
Morrison, Herbert (radio commentator), 48
Munich Airport, 12

Napier Lion engine, 138
National Laboratory of Psychical Research, 36, 40
Nazis, 45–7
Ndola, 122–8
Ndolo, 126–7
New York Herald Tribune, 66
New Zealand, 87, 89
Nicholls, Capt., 11
Nicosia, 139–40
Nimitz, Adm., 44–5
Noonan, Fred, 65–8
Nord guided missile, 122
Nordstern, 46
Northolt, 136–7
Norway, Nevil Shute, *see* Shute, Nevil

Oakland, Calif., 65–6, 89
O'Dea, Lt., 74
Omdal (pilot), 61
One-Eleven aircraft, 14
O'Neill, Sqn. Ldr., 30, 33, 35
Ontario, USS, 66
O'Reilly, Bill, 81
Orkun (Polish destroyer), 131
Oxford aircraft, *see* Airspeed

P. & O. shipping line, 61
Pacific Ocean, 52, 65–7, 79, 86, 97–8
Palstra, Sqn. Ldr., 30, 35
Pan American World Airways, 49, 52, 67, 95–8
Parapsychology Department, Duke University, 16

Parapsychology Foundation, New York, 44
Pardoe University, 65
Parke, Lt. Wilfred, 120
Patron, Monsieur, 33
Pentland, Capt. Charles, 108–9, 117, 119–20
Percival:
  Gull, 89
  Mew Gull, 94
Perkins, Marion, 64
Perkins, Sir Robert, 116
Pethybridge (engineer), 89, 94
Poix, 32
Portal, Marshal of the RAF, 53–6
Portchartrain (US weather ship), 97–8
Postal Monoplane, *see* Fairey
Potez Fouga, 122, 123
Powder Puff Derby, 65
Pratt and Witney engine, 93
Prchal (Czech pilot), 130–1
Prestwick, 70–1, 139
Price, Harry, 36, 43
Priestly, J.B., 133
Profumo, John, 116
Project Blue Book, 145–6
Project Flying Saucer, 144
Pruss, Capt., 47–50
Putnam, George Palmer, 64–5, 69, 73

Qantas Empire Airways, 80, 87, 94
Queen Juliana (KLM Constellation), 139

R101, etc., *see* airships
Rabouille, Monsieur, 33
Radcliffe (rigger), 29
Radcliffe, Mrs (rigger's wife), 36, 38
RAF, 23, 61, 81, 90, 104–5, 115, 130, 135–6, 139, 141, 143
RAF Transport Command, 130
Rauch, Kathie, 46–7
Rawlinson, Patricia, 110–11
Rhine, J.B., 16, 65
Richmond, Col., 21–3, 27, 29, 35, 40
Richmond, Mrs, 42
Robertson, Morgan, 15; *Futility*, 15
Rodley, Capt., 107–8
Rolls-Royce engine, 12
Rome, 104–7, 112–14, 120
Roosevelt, Franklin D., 44, 53, 67
Rosendahl, Capt., 48
Royal Aero Club, 81

Royal Aeronautical Society, 81
Royal Aircraft Establishment, *see* Farnborough
Royal Airship Works, *see* Cardington
Royal Australian Air Force, 30, 91

St Elmo's fire, 50
St Exupéry, Antoine de, 93, 157
St Raphael, 20, 61
Savory (R101 crewman), 34
Schmerling, Max, 46
Schneider Trophy, 72
Scott, C.W.A., 86
Scott, Maj. G.H., 21, 23, 25–32, 35, 40–1
Scott, Mrs G.H., 42
Segrave Memorial Medal, 81
Seymour, Ken, 140
Shackleton, *see* Avro
Shortridge, Travis, 80, 82–3, 85–6, 93–4
Shortridge, Mrs Travis, 83
Shortstown, 23, 42
Shute, Nevil, 22, 74
Sikorski, Gen., 129–32
Sikorsky amphibian, 61
Simon, Sir John, 38–44
Simson, Sir Henry, 30
Skyship 500, *see* airships
Skyways Operations, 11
Smith, Elinor, 64
Sonter, Tom, 91–2
Sopwith Pup, 79
Sopwith, Tom, 133
South Cerney Training School, 70
Southern Cloud, 73, 81–94
Southern Cross (formerly *Detroiter*), 80–2, 84, 87, 79, 94
Southern Cross Junior, 81, 86, 88, 94
Southern Cross Minor, 87–8, 91, 93–4
Southern Moon, 82
Southern Star, 84–5
Southern Sun, 73, 87
Soviet Air Force, 145
Spanner, E.F., 43; *The Tragedy of the R101*, 43
Spooner, Winifred, 28, 30
Stanstead, 11
Star Ariel, 12, 20
Star Tiger, 12, 20
Stead, W.T., 135
Stella Australis, 89, 94
Stewart, Bill, 146

Stiles, Sqn. Ldr., 139
Stinson monoplane, 61–2
Stokes, Claire, 82, 92
Stokes, Mrs (mother of Claire), 92
Stolyarov, Maj. Gen., 145
Stratocruiser, *see* Boeing
Stulz (pilot), 61, 64
Swissair, 14
Sydney Morning Herald, 82–3, 90
'synchronicity', 101

Tait, Sir Victor, 112
Taylor, P.G., 89, 91
Teacher (brother-in-law of Irwin), 41
Teacher, Olive, *see* Irwin, Olive
Teed, Maj., 38–9, 41–2, 44
telepathy, 15–16
Tempest, HMS, 36
Terzi, Francesca Lana, 142, 151
Thain, Capt., 12–13
Tholen (Dutch psychic), 139
Thomson, Lord (Christopher Birdwood),
    21–2, 24, 26–9, 30–1, 35, 37–8,
    40–5
Thomson, Miss (Lord Thomson's sister),
    42
Tillier, Louis, 33
Time, 15, 133, 162
Times, The, 13, 48, 118, 124, 128, 137–8
Titan, 15
Titanic, 15
Topcott, Miss (medium), 149
Transair, 123, 126
Trenchard, Lord, 42, 137
Trent (farmer), 145–6
Tribune, HMS, 36
Triplex glass, 21–2
Tripoli, 57–8
Tshombe, Gen. Moise, 121–3, 125,
    127–8
Tudor, *see* Avro

UFOs (unidentified flying objects), 144–8
Ulm, Charles, 73, 80, 82–3, 85–6, 89,
    94
United Nations, 121–8
UNOC (United Nations Organization in
    the Congo), 127
United Services Club, 150
United States Air Force, 126, 146–7

United States Navy, 67, 69
University of Colorado, 145
Upavon, 61
'Uvani', 36–7, 44; *see* also Irwin, Flt. Lt.
    Harry

Van Dyck (Dutch flyer), 89
Vansittart, Lord, 76
VC10, *see* British Aircraft Co.
Verne, Jules, 52, 133; *From Earth to
    Moon*, 52
Vildebeeste bomber, 90
Villiers, Maj. Oliver, 28, 30–1, 38–40, 43
Von Huenefeld, Baron Gunther, 61
'voodoo' aircraft, 105

Wackett, Sir Lawrence, 93
Wakefield, Lord, 73
Wallis, Barnes, 21, 38
Walton, First Officer, 11
Waugh, Bruce, 75, 77
Wellbrooke, Harry, 49
Wells, H.G., 133; *The Time Machine*, 133
West Australian Airways, 79
Westland Widgeon, 80
Whiting, James Max, 149–50
Whiting, Mr (father of Max Whiting),
    149
Whiting, Muriel, *see* Dowding, Lady
Widgeon, *see* Westland Widgeon
Wilkins, Sir George, 80
Williams, Cons., 85
Wilson, Charles, 53
Witkin, H.A., 119
Woolf, Virginia, 37
World War I, 29, 62–3, 79, 143
World War II, 45, 73–4, 104, 129, 140,
    143–4, 148
Wrangler, HMS, 114
Wright, Orville and Wilbur, 133, 151

Yoke Peter, etc., *see* De Havilland
    Comet
York aircraft, *see* Avro

Zeigarnik Effect, 49
Zeppelin, *see* airships
Zeppelin, Count, 45
Zeppelin Reederei, Frankfurt, 45–6
Zichy, Count, 71